Mohamed Elzagheid
Polymers

Also of Interest

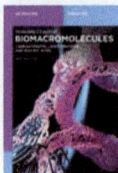

Biomacromolecules.
Carbohydrates, Lipids, Proteins and Nucleic Acids
Mohamed Elzagheid, 2025
ISBN 978-3-11-158298-6, e-ISBN (PDF) 978-3-11-158327-3

Organic Chemistry: 25 Must-Know Classes of Organic Compounds
Mohamed Elzagheid, 2024
ISBN 978-3-11-138199-2, e-ISBN 978-3-11-138275-3

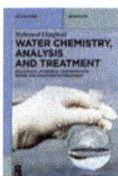

Water Chemistry, Analysis and Treatment.
Pollutants, Microbial Contaminants, Water and Wastewater Treatment
Mohamed Elzagheid, 2024
ISBN 978-3-11-133242-0, e-ISBN 978-3-11-133246-8

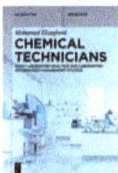

Chemical Technicians.
Good Laboratory Practice and Laboratory Information Management Systems
Mohamed Elzagheid, 2023
ISBN 978-3-11-119110-2, e-ISBN 978-3-11-119149-2

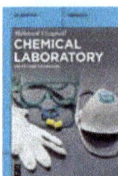

Chemical Laboratory.
Safety and Techniques
Mohamed Elzagheid, 2022
ISBN 978-3-11-077911-0, e-ISBN 978-3-11-077912-7

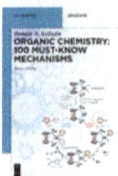

Organic Chemistry: 100 Must-Know Mechanisms
Roman Valiulin, 2023
ISBN 978-3-11-078682-8, e-ISBN 978-3-11-078683-5

Mohamed Elzagheid

Polymers

Chemistry, Morphology, Characterization, Processing,
Technology and Recycling

2nd, expanded edition

DE GRUYTER

Author
Prof. Dr. Mohamed Elzagheid
Royal Commission for Jubail and Yanbu
Jubail Industrial College
Jubail Industrial City, Saudi Arabia
and
Center for Research and Strategic Studies
Libyan Authority for Scientific Research
Tripoli, Libya
melzagheid@gmail.com

ISBN 978-3-11-158565-9
e-ISBN (PDF) 978-3-11-158573-4
e-ISBN (EPUB) 978-3-11-158605-2

Library of Congress Control Number: 2025937001

Bibliographic information published by the Deutsche Nationalbibliothek
The Deutsche Nationalbibliothek lists this publication in the Deutsche Nationalbibliografie; detailed
bibliographic data are available on the Internet at http://dnb.dnb.de.

© 2025 Walter de Gruyter GmbH, Berlin/Boston, Genthiner Straße 13, 10785 Berlin
Cover image: Irina Vodneva/iStock/Getty Images Plus
Typesetting: Integra Software Services Pvt. Ltd.

www.degruyter.com
Questions about General Product Safety Regulation:
productsafety@degruyterbrill.com

This work is dedicated to the fascinating world of polymers, where chemistry meets ingenuity and innovation shapes the future.

Preface

Polymers are vital for modern living, impacting every corner of our everyday lives, from the materials we use to the technology we rely on. Their accessibility and attainability make them suitable for a wide range of applications, including industry, environmental sustainability, and medicine.

This book addresses the chemistry, structure, processing, characterization, technology, and key issues of polymers, including sustainability and recycling. **Chapter 1** covers the basic concepts of polymers and their historical evolution, beginning with natural polymers such as silk, rubber, and wool, and concluding with synthetic materials such as PS and PVC that have significantly impacted technology. Polymer classification based on their origin, structure, and features, which allows for more in-depth knowledge, is also discussed. The chemistry of polymerization is explored in **Chapter 2**, with different polymerization techniques, including addition, condensation, and ring-opening polymerization, being discussed. This chapter also covers copolymerization and polymer modifications, with a special focus on biopolymers and green chemistry approaches, and concludes with a brief discussion on organic polymers and inorganic polymers. Polymer morphology is a critical area of study, and **Chapter 3** addresses the relationship between molecular structure and physical properties. From crystallinity to polymer blends, the chapter provides a comprehensive exploration of how polymers are organized at the microscopic level and how their morphology influences their performance. To fully understand polymers, it is essential to examine their characteristics through advanced techniques. **Chapter 4** delves into polymer characterization, including methods to measure molecular weight distribution, thermal properties, mechanical behavior, and rheology. Spectroscopic, calorimetric, and microscopic techniques such as FTIR, NMR, XRD, DSC, SEM, AFM, and TEM allow for detailed exploration of the molecular and structural aspects of polymers. The processing of polymers is essential for creating functional materials, and **Chapter 5** discusses various processing technologies such as extrusion, injection molding, and 3D printing. It also highlights the integration of nanotechnology into polymer processing to enhance material properties. **Chapter 6** delves into cutting-edge polymers, discussing their types and applications. **Chapter 7** focuses primarily on polymer product design and applications. It examines how to select and shape polymers for a variety of uses, ranging from consumer products to industrial components. This chapter also explores the practical applications of polymer technology, including its use in the automotive, aerospace, electronics, and healthcare industries. The development of green polymers and sustainable processes receives considerable attention as the polymer industry places a greater emphasis on environmental responsibility. **Chapter 8** explores polymer recycling, upcycling, and downcycling technologies, as well as the future of biopolymer repurposing. It addresses the challenges and opportunities in managing polymer waste. Finally, **Chapter 9** briefly discusses the environmental and societal impact of polymers and presents solutions and strategies to address these issues.

https://doi.org/10.1515/9783111585734-202

This book is designed as a comprehensive resource for students, researchers, and professionals interested in the ever-evolving field of polymer science. It seeks to provide both foundational knowledge and advanced insights, equipping readers with the tools to understand and innovate in the world of polymers.

Mohamed Ibrahim Elzagheid, Chemistry Professor
Waterloo, Ontario, Canada
2025

Acknowledgment

First and foremost, I want to thank my family for their everlasting support, inspiration, and love throughout my academic career. Their unwavering belief in me, even during difficult times, has provided me with strength and determination. I will be eternally thankful for their patience and understanding, which have allowed me to accomplish this goal.

I would also like to thank my colleagues at Jubail Industrial College and the Libyan Authority for Scientific Research. Their encouragement and support have been immensely beneficial. I am also grateful for the opportunity to contribute to the scientific community by collecting a significant number of textbooks for the library, which will serve as a valuable resource for students and academics in the future.

Last but not least, I'd like to thank the De Gruyter publishing team for their invaluable support during this process. I'm grateful to Ute Skambraks, Helene Chavaroche, Nadja Schedensack, and Chandhini Magesh for their expertise, dedication, and assistance in completing this project.

https://doi.org/10.1515/9783111585734-203

Contents

The Author

Professor Mohamed Elzagheid is a chemist, scientist, author, and teacher. In addition to his chemistry research, he specializes in chemical education, academic advising, curriculum innovation, vocational education, and higher education administration.

Throughout his 30-year career at Turku University in Finland, McGill University, SynPrep Inc. in Montreal, Canada, and Jubail Industrial College in Saudi Arabia, he has been directly and indirectly involved in the education of laboratory technicians and chemists, as well as in supervising numerous undergraduate and graduate students.

He has made significant contributions to various short- and long-term training programs for Saudi enterprises, in addition to teaching a diverse variety of chemistry and chemistry-related courses at all levels. These courses include basic and advanced organic chemistry, polymer chemistry, macromolecular chemistry, biochemistry, laboratory techniques, safety in chemical laboratories, technician responsibilities, geology and petroleum geology, and water-wastewater analysis and treatment.

Dr. Elzagheid is the author of eight textbooks: *Introductory Organic Chemistry; Thoughts on Organic Chemistry; Macromolecular Chemistry: Natural and Synthetic Polymers; Chemical Laboratory Safety, Techniques; Chemical Technicians: Good Laboratory Practice and Laboratory Information Management Systems; Water Chemistry, Analysis and Treatment: Pollutants, Microbial Contaminants, Water and Wastewater Treatment; Organic Chemistry: 25 Must-Know Classes of Organic Compounds;* and *Biomacromolecules: Carbohydrates, Lipids, Proteins, and Nucleic Acids.*

His work at Turku University in Finland, McGill University in Canada, and Jubail Industrial College in the Kingdom of Saudi Arabia has helped him build a solid reputation in both chemistry and chemical education, as demonstrated by his research papers and publications.

https://doi.org/10.1515/9783111585734-205

Chapter 1
Introduction to Polymers

1.1 Polymers History

A polymer, or macromolecule, is any material composed of many repeating subunits, building blocks, or monomers. Natural and synthetic polymers not only serve a wide range of important functions in our daily lives but are also essential for biological structure and function. They range from simple synthetic polymers like polyethylene, polypropylene, and polystyrene to naturally occurring biopolymers such as proteins, polysaccharides, and nucleic acids (DNA and RNA). Both natural and man-made polymers are formed by polymerizing small units known as monomers. They have large molecular masses and exhibit distinct physical properties such as toughness, high elasticity, and viscoelasticity, which distinguish them from small molecules.

Although most synthetic polymers were not discovered until the nineteenth and twentieth centuries, natural polymers have been used for thousands of years. For example, rubber (Figure 1.1) has been obtained from the latex of rubber trees by the indigenous peoples of Central and South America. Cellulose, another natural polymer, has been used in textiles since ancient times – Egyptians, for instance, used flax to manufacture linen (Figure 1.2). Animal-derived polymers include silk and wool. Silk is produced by silkworms (Figure 1.3), while wool is obtained from sheep (Figure 1.4). Both have been used in fabrics and textiles for millennia.

Figure 1.1: The latex of rubber trees (image credit: https://ecoterrabeds.com/blogs/eco-terras-healthy-sleep-blog/where-does-latex-come-from).

https://doi.org/10.1515/9783111585734-001

Figure 1.2: Flax and linen fibers (image credit: https://www.textilecoach.net/post/flax-or-linen-fiber).

Figure 1.3: Silkworms produce silk (image credit: Timekeep/Shutterstock. https://www.discovermagazine. com/planet-earth/silk-making-is-an-ancient-practice-that-presents-an-ethical-dilemma).

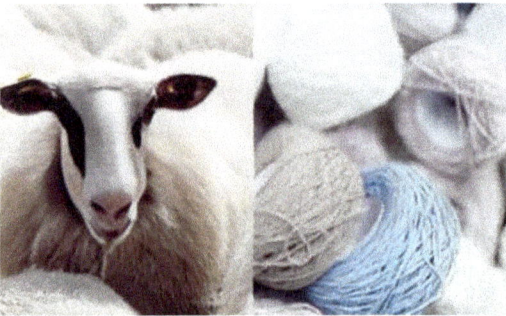

Figure 1.4: Sheep produce wool (image credit: https://www.linkedin.com/pulse/wool-where-comfort-meets-style-merchantbay/).

1.2 The Origin of Polymer Science

1.2.1 The Term "Polymer"

Jöns Jacob Berzelius, a Swedish scientist, used the term "polymer" in 1833 to characterize compounds composed of repeated structural units, now known as monomers. His hypothesis was based on his studies of organic chemicals and how basic molecules interact with larger, more complex structures. However, his notion of polymers differed from current understanding.

It wasn't until the twentieth century that scientists gained a better understanding of polymers as long-chain compounds made up of repeating units, either naturally occurring or manmade. During this time, synthetic polymers advanced rapidly, thanks in large part to the efforts of pioneers such as Leo Baekeland, who invented Bakelite, the first synthetic plastic, and Hermann Staudinger, whose 1920s research laid the framework for macromolecular chemistry.

Polymer is derived from the Greek words *poly* ("many") and *meros* ("part" or "unit"), which refer to a structure made up of several repeating components. Polymers can exist in both natural and manmade forms. Natural polymers include DNA, proteins, starch, and natural rubber, whereas synthetic polymers consist of plastics, nylon, and synthetic rubber.

1.2.2 The Discovery of Natural Rubber

The discovery of natural rubber is an important chapter in both scientific and industrial history. Indigenous peoples in South America were the first to use rubber, particularly in the Amazon Basin, where they collected the milky sap from rubber trees (Hevea brasiliensis) to create waterproof goods like footwear and clothing.

The word "rubber" is derived from its capacity to "rub out" pencil markings. Although European scientists began examining it in the seventeenth century, rubber's actual economic potential was not recognized until the nineteenth century, thanks in large part to Charles Goodyear's discovery of the vulcanization process in 1839. Vulcanization increased the durability, elasticity, and weather resistance of rubber, allowing it to be utilized in a variety of applications, such as tires and waterproof apparel.

During the nineteenth-century rubber boom, the demand for rubber surged, driving the industry's expansion. A significant turning point occurred in 1876 when British colonial officer Henry Wickham secretly transported rubber tree seeds from Brazil to Southeast Asia. This act led to the development of extensive rubber plantations in Malaysia and Indonesia. The global cultivation of rubber helped meet the growing demand, making it an essential material in various industries, including automotive and medical applications.

1.2.3 The Discovery of Cellulose Derivatives

The nineteenth century marked a breakthrough in cellulose derivatives as research-ers explored the possible commercial uses of cellulose, a naturally occurring polymer found in plant cell walls. In 1838, Anselme Payen, a French scientist, was the first to identify cellulose as a distinct molecule capable of further chemical modification. In 1855, British scientist Thomas Hill successfully esterified cellulose by treating it with acetic acid to produce cellulose acetate, marking the first significant advance in cellu-lose derivatives. The discovery that cellulose could be chemically modified to generate a variety of useful molecules sparked the creation of cellulose derivatives.

In the years that followed, numerous additional cellulose derivatives were pro-duced, including cellulose nitrate (discovered in 1846 by Christian Friedrich Schön-bein), which led to the creation of celluloid, one of the earliest synthetic polymers. The subsequent discovery and production of other cellulose derivatives, such as meth-ylcellulose and carboxymethylcellulose, transformed industries such as textiles, food, and pharmaceuticals, as these derivatives outperformed the original cellulose in terms of solubility, flexibility, and biodegradability. Cellulose acetate, in particular, has found extensive use in the textile industry as a precursor for synthetic fibers and in the photographic film sector. These advances paved the way for the larger area of polymer chemistry, as well as the creation of various additional synthetic polymers and materials, emphasizing cellulose as a useful raw resource for modern industrial applications.

1.2.4 The Invention of the First Synthetic Plastic

In 1869, John Wesley Hyatt developed the first synthetic polymer in response to a $10,000 award offered by a New York corporation for an alternative to ivory. The expanding popularity of billiards had greatly boosted the demand for natural ivory, which was obtained by hunting wild elephants. Hyatt discovered that by processing cellulose from cotton fibers with camphor, he could create a plastic that could be shaped into various forms and simulate materials like tortoiseshell, horn, linen, and ivory. This achievement was a turning point in manufacturing, freeing production from the constraints of naturally available resources.

1.2.5 The History of Polystyrene Synthesis

The journey of polystyrene synthesis begins in 1839, when German chemist Henri Vic-tor Regnault created styrene, a liquid hydrocarbon derived from turpentine. Styrene turned out to be an important monomer that could polymerize into a solid substance, but its full potential wasn't recognized at the time. In 1928, German chemists Irenäus

T. K. and Ludwig discovered a procedure known as radical polymerization to polymerize styrene. This was a significant step forward in synthetic polymer synthesis, demonstrating that styrene could be polymerized into a solid, crystalline form.

The German company IG Farben began commercial manufacturing of polystyrene in 1931. They developed a method for producing polystyrene by polymerizing styrene monomers in bulk, which was extremely important as it allowed for the production of polystyrene in large quantities. Dow Chemical in the United States began manufacturing polystyrene in 1937, introducing it to the mainstream market. The development of expanded polystyrene (EPS) in the 1940s marked another significant milestone. By incorporating a foaming agent into the polymerization process, a lightweight, foam variant of polystyrene was created, which became widely used for insulation and packaging materials (e.g., Styrofoam). By the 1960s, polystyrene had become one of the world's most widely produced polymers, with a broad range of applications in industries, including packaging and consumer goods. Polystyrene was further enhanced over the years, and styrene-based copolymers such as acrylonitrile butadiene styrene (ABS) were developed to improve the material's properties for specific uses, such as greater impact resistance and heat stability. Although polystyrene is still widely used today, its negative environmental effects, particularly in the case of Styrofoam, have led to increased calls for recycling and the development of alternative materials. Despite these concerns, its history represents a pivotal moment in the creation of synthetic polymers for industrial and consumer applications, as well as a notable advancement in polymer chemistry.

1.2.6 The Polymer Boom

The Polymer Boom describes the rapid expansion of the polymer industry in the twentieth century, particularly following World War II, when polymers evolved from experimental materials into indispensable components across various sectors. This surge in polymer research and production reshaped manufacturing, driving advancements in packaging, automotive engineering, electronics, medical devices, and consumer goods. The roots of this boom trace back to key discoveries in the nineteenth and early twentieth centuries, such as the invention of Bakelite in 1907 by Leo Baekeland and the development of nylon in 1935 by Wallace Carothers at DuPont. These pioneering synthetic polymers demonstrated the potential to create strong, adaptable materials that could be mass-produced. However, the true polymer boom gained momentum in the 1940s and 1950s, spurred by new polymerization techniques, rising industrial demand, and innovations driven by wartime needs. World War II sparked a significant drive to develop synthetic materials as alternatives to scarce natural resources. For instance, the shortage of natural rubber led to the creation of butadiene rubber, which had a lasting impact on the automotive industry and beyond. Likewise, nylon, originally developed for military applications such as parachutes and ropes, gained widespread popularity in the consumer market after the war, particularly in textiles. In the post-war era, the commercialization of new synthetic plastics surged, including polyethylene (PE), polypropylene (PP), polystyrene (PS), and polyvinyl chloride (PVC). These materials were lightweight, cost-effective, and highly moldable, making them ideal for large-scale manufacturing and revolutionizing numerous industries.

During the 1950s and 1960s, the introduction of high-density polyethylene (HDPE) and acrylonitrile butadiene styrene (ABS) further accelerated the growth of the polymer industry. As these advanced materials became more accessible, they were incorporated into a wide range of applications, including packaging, automotive components, toys, and electronic devices.

The rapid expansion of the polymer industry has had significant economic implications, fostering new markets and creating employment opportunities in manufacturing, chemical production, and research and development. The increasing demand for polymers in a fast-growing global economy has spurred advancements in polymerization techniques, leading to more efficient and cost-effective mass production. The introduction of additives and copolymers has enabled further customization, enhancing properties such as durability, flexibility, and resistance to heat and chemicals.

By the 1970s and 1980s, concerns over plastic waste led to a growing emphasis on recycling and sustainability. This prompted efforts to develop more eco-friendly and recyclable polymers. As environmental awareness increased, innovation in polymer science continued, giving rise to materials such as biodegradable plastics and specialized polymers for medical applications.

In conclusion, the polymer boom marked a transformative era in materials science, driven by groundbreaking discoveries and innovations in synthetic polymers. The widespread adoption of these materials revolutionized industries, manufacturing, and consumer products, shaping the modern world in profound ways.

1.2.7 The Development of Engineering Plastics

Engineering plastics represent a tremendous advancement in material science, offering high-performance alternatives to conventional metals, ceramics, and wood. Engineering plastics are synthetic polymers with excellent mechanical qualities such as strength, durability, heat resistance, chemical resistance, and dimensional stability. Engineering plastics are utilized in more demanding applications, such as automotive, aerospace, electronics, and industrial machinery, as opposed to commodity plastics, which are often used in everyday items such as bags, containers, and packaging. Their evolution is a tale of invention, scientific advancement, and industrial need.

The roots of engineering plastics trace back to the early twentieth century, with key developments in polymer chemistry and materials science. The first synthetic polymer, Bakelite, invented by Leo Baekeland in 1907, is often considered the precursor to modern engineering plastics. While Bakelite was primarily a thermosetting plastic used for electrical insulators and household items, its success demonstrated the potential for plastics to replace traditional materials. The true rise of engineering plastics began in the mid-twentieth century, driven by the need for materials that could withstand the growing demands of industries such as automotive, aerospace, and electronics.

The key developments in engineering plastics started in the 1940s and 1950s. Significant progress was made in developing polymers with enhanced thermal stability, strength, and resistance to chemicals. A few landmark materials that emerged during this time are:

- Polycarbonate (PC): In the 1950s, polycarbonate was introduced by General Electric Plastics. Polycarbonate's ability to withstand high temperatures and its inherent strength led to its adoption in many demanding applications, ranging from automotive parts to medical devices.
- Polyamide (PA): Polyamides (such as nylon) gained prominence in the 1950s and 1960s due to their excellent mechanical properties, wear resistance, and heat tolerance. These materials became important in applications such as automotive parts, electrical connectors, and textiles.
- Polyphenylene oxide (PPO): Developed in the late 1950s, PPO is a high-performance polymer known for its excellent electrical insulating properties, high-temperature stability, and resistance to chemicals. It has become widely used in the manufacturing of electrical components and automotive parts.
- Polyether ether ketone (PEEK): In the 1980s, PEEK was introduced as one of the most advanced engineering plastics, known for its high strength, resistance to heat (up to 250 °C), and excellent chemical resistance. PEEK became widely used in applications such as aerospace, medical implants, and automotive components.

Engineering plastics have played a significant role in modern industries. The development of engineering plastics revolutionized various industries by providing materials that are lightweight, durable, and resistant to extreme conditions. Their ability to replace metals in many applications has resulted in cost savings, weight reductions, and improved performance. Table 1.1 shows the areas where engineering plastics have had a significant impact.

While engineering plastics have become crucial in modern manufacturing, they are not without their challenges. Issues such as environmental impact, recycling, and biodegradability have become significant concerns, especially as the use of plastics continues to rise. In response, research and researchers are focusing on developing bio-based engineering plastics that are more sustainable and easier to recycle, as well as on enhancing the properties of engineering plastics, allowing them to perform even better in extreme environments.

Table 1.1: Areas where engineering plastics have made a significant impact.

Area	Impact
Automotive industry	Engineering plastics have become integral to the automotive sector, replacing metals in applications such as bumpers, dashboard components, fuel systems, and electrical connectors. The use of plastics in automotive parts reduces vehicle weight, which improves fuel efficiency and performance.
Aerospace	Engineering plastics are used in applications where lightweight materials with high strength and heat resistance are crucial. They are utilized in engine components, airframes, and seals, where they can withstand high temperatures and extreme conditions.
Electronics	Engineering plastics have excellent insulating properties and resistance to high temperatures, which make them ideal for manufacturing electronic components.
Medical devices	High-performance engineering plastics are utilized in medical devices, such as implants, surgical instruments, and drug delivery systems, because of their biocompatibility, strength, and resistance to sterilization processes.

1.2.8 The Development of Polymer Blends and Composites

The history of polymer blends and composites dates back to the twentieth century, paralleling the rise of synthetic polymers and the quest for materials with enhanced properties. Their development stemmed from the need to combine the advantages of different polymers or reinforce polymers with other materials, resulting in improved strength, durability, heat resistance, and other desirable characteristics.

Polymer blending originated in the 1920s and 1930s as a result of early polymer science investigations. However, the technique of mixing different polymers gained popularity in the 1940s and 1950s, when academics and industry experts realized that blending could outperform single-polymer systems. Initially, most polymer blends were immiscible, meaning they did not mix at the molecular level but instead formed different phases within the material, influencing its overall properties. A significant early example of polymer blending was the combination of natural and synthetic rubber during World War II. Due to natural rubber shortages, synthetic rubbers were increasingly used and blended with natural rubber to enhance mechanical properties such as wear resistance and flexibility to meet wartime demands in tire and rubber product manufacturing.

As polymer technology improved, notably when Wallace Carothers of DuPont invented nylon in the 1930s, researchers began experimenting with combining various synthetic polymers. However, it was not until the 1950s and 1960s that the true potential of polymer mixtures became apparent. These decades saw the advent of NORYL, a poly(styrene) (PS) and poly(phenylene oxide) (PPO) miscible mix created in 1966 by

General Electric Plastics and later acquired by SABIC. These mixes outperformed individual polymers in terms of processing, toughness, and optical clarity.

The development of polymer composites also took off in the late 1940s and 1950s, spurred by the need for stronger, lighter, and more flexible materials in industries such as aircraft and automobile manufacturing. Composites, unlike polymer blends, are made by embedding reinforcing elements (such as fibers or particles) into a polymer matrix, greatly improving mechanical characteristics, thermal stability, and durability. Fiberglass was introduced in the 1940s, marking a significant early accomplishment in polymer composite research. Fiberglass, which is made up of glass fibers embedded in a polymer matrix, often polyester or epoxy, has a high strength-to-weight ratio, great corrosion resistance, and effective electrical insulating qualities. Fiberglass was widely employed in the manufacture of boats, automobile parts, aircraft components, and insulating materials. This was the first significant use of fiber-reinforced composites, which would subsequently become the foundation of advanced composites in sectors that required high-performance materials. Carbon fibers were also investigated in the 1950s and 1960s because of their excellent strength and stiffness-to-weight ratios. Carbon fiber-reinforced composites (CFRPs) gained popularity in aerospace and defense applications by the 1970s, notably in the production of airplanes, missiles, and satellites. The impact of key early polymer blends and composites is outlined in Table 1.2.

Table 1.2: The impact of key early polymer blends and composites.

Polymer material	Impact
Bakelite	Although Bakelite is not a polymer blend or composite, its manufacture as the first polymer led to the development of advanced polymeric materials.
Nylon	Wallace Carothers of DuPont invented nylon, which was a major achievement in polymer science. Although nylon is not a blend or composite, its success laid the groundwork for future advances in polymer research, such as the creation of polymer blends and composites.
Synthetic rubber blends	During World War II, the shortage of natural rubber led to the blending of synthetic rubber, such as butadiene rubber, with natural rubber to create a more durable and flexible material for tires and military applications. This early use of polymer blends highlighted the potential benefits of combining different polymers.
Fiberglass	The development of polymer composites took a significant step forward with the introduction of fiberglass reinforced with polyester or epoxy resins. The integration of glass fibers with a polymer matrix resulted in lightweight, strong, and corrosion-resistant materials, making them especially valuable in the automotive and aerospace industries.

Table 1.2 (continued)

Polymer material	Impact
Carbon fiber	Carbon fibers were originally developed for aerospace applications because of their outstanding strength and heat resistance. When incorporated into a polymer matrix, they produce composite materials with a high strength-to-weight ratio, leading to the creation of carbon fiber-reinforced plastics (CFRPs). These innovative composites have revolutionized the aerospace, automotive, and sports equipment industries.
Epoxy resins	The development of epoxy resins as a matrix material for fiber-reinforced composites was a breakthrough. Epoxy resins offered excellent adhesion to fibers like glass and carbon, leading to the widespread use of fiber-reinforced composites in high-performance applications, particularly in the aerospace and marine industries.

When it comes to challenges and limitations in the early development of polymer blends and composites, several obstacles were encountered during the initial years. These challenges can be summarized as follows:

- Poor compatibility between different polymers.
- Challenges in processing.
- The immiscibility of certain polymers.
- Dispersion of fibers within the matrix.
- Adhesion between the polymer matrix and the reinforcing fibers.

1.2.9 The Development Biodegradable Polymers

The development of biodegradable polymers is deeply connected to the advancement of polymer science, particularly the study of natural and synthetic materials that can break down in the environment. The concept of biodegradable polymer materials, which microorganisms degrade into water, carbon dioxide, and biomass, gained significant attention in the mid-twentieth century as awareness of plastic waste and its environmental impact increased. However, biodegradable materials themselves are not a recent innovation. Natural biodegradable polymers, including starch, cellulose, and protein-based substances like gelatin and silk, have been in use for thousands of years.

These natural polymers are biodegradable because their molecular structures enable microorganisms to break them down efficiently in the environment. Cellulose, for instance, is a widely abundant polymer found in plant cell walls. Highly biodegradable, it has traditionally been used in paper, textiles, and packaging materials. Similarly, starch, another naturally occurring polymer, has been utilized for centuries. Its biodegradability makes it an excellent choice for various applications, including packaging, as it naturally decomposes when exposed to moisture and microbial activity.

Protein-based materials, such as gelatin derived from animal collagen, have historically been used in packaging and medical applications. These materials are naturally biodegradable, as microorganisms break them down by digesting proteins. Similarly, natural rubber has been widely used for various products due to its biodegradability. Although rubber did not achieve widespread commercial use until the nineteenth century, it has long been recognized for its ability to degrade in the environment, unlike synthetic rubbers and plastics.

The development of synthetic biodegradable polymers began in the mid-twentieth century in response to growing concerns about plastic waste and its environmental impact. One of the earliest discoveries in this field was the identification of polylactic acid (PLA) in the 1930s. PLA, which is made from renewable sources like maize starch and sugarcane, was first recognized for its biodegradability. However, it wasn't until environmental concerns grew in the 1970s and 1980s that PLA was widely recognized as a viable alternative to traditional petroleum-based plastics. Commercial manufacturing of PLA began in the 1980s when corporations improved polymerization techniques to enable large-scale production.

Another important milestone came with the development of polyhydroxyalkanoates (PHA), a family of biodegradable plastics produced by bacteria through the fermentation of sugars or lipids. PHAs were discovered in the 1920s by scientists studying bacterial metabolism. However, it was in the 1970s and 1980s that PHA production for commercial use became more feasible. These materials are fully biodegradable, breaking down through microbial activity in the environment. They have been considered alternatives to petroleum-based plastics, particularly for applications such as packaging, agricultural films, and medical products.

Starch-based biodegradable polymers were also explored in the 1970s as part of efforts to find renewable, biodegradable materials that could replace petroleum-based plastics. Starch was modified chemically or physically to improve its properties, and products such as starch-based packaging and plastic films began to appear on the market in the 1980s. The key advantage of starch-based polymers is their ability to break down easily in the environment due to the natural biodegradation of starch. Starch-based biodegradable plastics are particularly suitable for single-use products, such as food packaging and disposable cutlery, because of their biodegradability and combustibility. However, these materials generally require modifications to improve their mechanical properties and processing characteristics, as natural starch alone is not a suitable substitute for more durable plastic products.

The commercialization of biodegradable polymers truly accelerated in the 1990s and 2000s, driven by growing environmental awareness, governmental pressure to reduce plastic waste, and consumer demand for more sustainable products. During this time, biodegradable plastics began to see widespread use in applications such as food packaging, biomedical devices, and agricultural films. In the 1990s, a growing recognition of the environmental impact of plastic pollution, especially in oceans and landfills, spurred the development of alternatives. Compostable plastics, often made from cornstarch or PLA, gained traction in the marketplace. These materials were not only biodegradable but also compostable, meaning they could break down in controlled environments like industrial composting facilities. While significant progress has been made in developing biodegradable polymers, challenges related to cost, processing, and scaling up production still remain. However, the continued demand for sustainable materials is expected to drive further innovation in this field.

1.3 Chronological Developments of Commercial Polymers

Polymers are produced by industry in large quantities due to their commercial importance. Nowadays, a large number of these polymers are available and have many applications. Table 1.3 presents their chronological development.

Table 1.3: Chronological development of commercial polymers.

Years	Polymers
1839–1868	– Vulcanization of rubber
	– Cellulose esters
	– Nitration of cellulose
	– Ebonite (hard rubber)
	– Celluloid (plasticized cellulose nitrate)
1888–1907	– Pneumatic tires
	– Cellulose nitrate photographic films
	– Cuprammonium rayon fibers
	– Viscose rayon fibers
	– Poly(phenylene sulfide)
	– Glyptal polyesters
	– First tubeless tire
	– Phenol-formaldehyde resins (Bakelite)
1908–1928	– Cellulose acetate photographic fibers
	– Regenerated cellulose sheet (cellophane)
	– Poly(vinyl acetate)
	– Urea-formaldehyde resins
	– Cellulose nitrate
	– Cellulose acetate fibers
	– Alkyd
	– Poly(vinyl chloride)
	– Cellulose acetate sheets
	– Nylon

Table 1.3 (continued)

Years	Polymers
1929–1939	– Polysulfide synthetic elastomer
	– Polyethylene
	– Poly(methyl methacrylate)
	– Polychloroprene elastomer (neoprene)
	– Epoxy resins
	– Ethyl cellulose
	– Poly(vinyl butyral)
	– Polystyrene
	– Styrene-butadiene rubber (Buna-S) and styrene-acrylonitrile (Buna-N) copolymer elastomers
	– Melamine-formaldehyde resins
	– Nylon-6
	– Nitrile rubber
1940–1950	– Isobutylene-isoprene elastomer
	– Low-density polyethylene
	– Poly(ethylene terephthalate)
	– Butyl rubber
	– Unsaturated polyesters
	– Fluorocarbon resins (Teflon)
	– Silicones
	– Styrene-butadiene rubber
	– Polysulfide rubber
	– Acrylonitrile-butadiene-styrene copolymers
	– Cyanoacrylate
	– Polyester fibers
	– Polyacrylonitrile fibers
1952–1962	– High-impact polystyrene
	– Polycarbonates
	– Polyphenylene ether (polyphenylene oxide)
	– High-density polyethylene
	– Polypropylene
	– Polycarbonate
	– Poly(dihydroxymethylcyclohexyl terephthalate)
	– Ethylene-propylene monomer elastomers
	– Aromatic nylons (aramids)
	– Polyimide resins

Table 1.3 (continued)

Years	Polymers
1964–1982	– Poly(phenylene oxide)
	– Ionomers
	– Polysulfones
	– Styrene-butadiene block copolymers
	– Liquid crystals
	– Poly(butylene terephthalate)
	– Polyacetylene
	– Polyetherimide
1990–2000s	– Carbon nanotubes
	– Starch-based biodegradable polymers

1.4 Macromolecules vs Polymers

A macromolecule is a large polymeric or non-polymeric molecule with a high molecular mass. Polymers are large molecules with a high molecular weight, made up of small monomeric repeating units that are typically bonded or connected by covalent bonds.

Pectin, cellulose, polyamide, polycarbonate, and polyethylene terephthalate are macromolecules that are categorized as polymers since they are composed of repeated building blocks, or monomers. Similarly, proteins and nucleic acids are polymers due to their repeating sequences. On the other hand, lignin and montan wax are composed of a distinct set of molecules that are not repeating monomers; they are macromolecules but cannot be described as polymers. In general, the term "macromolecule" covers biomacromolecules (biopolymers) such as proteins, polysaccharides, nucleic acids, lipids, and synthetic polymers such as polyethylene oxide, polyethylene, polypropylene, and other synthetic polymers. Biomacromolecules are natural polymers that are found in nature, and some of their analogs can be made in the laboratory. Synthetic polymers are man-made polymers that are synthesized from small organic and inorganic molecules, and most of them are made from oil and gas products. Examples of selected macromolecules and polymers are shown in Figure 1.5.

1.5 Natural vs Synthetic Polymers

Biomacromolecules, also known as biopolymers or natural polymers, are composed of carbohydrates (sugars), proteins (peptides and polypeptides), and nucleic acids (DNA and RNA). Their monomers consist of monosaccharides, amino acids, and nucleotides. Synthetic polymers are derived from small molecules and are classified

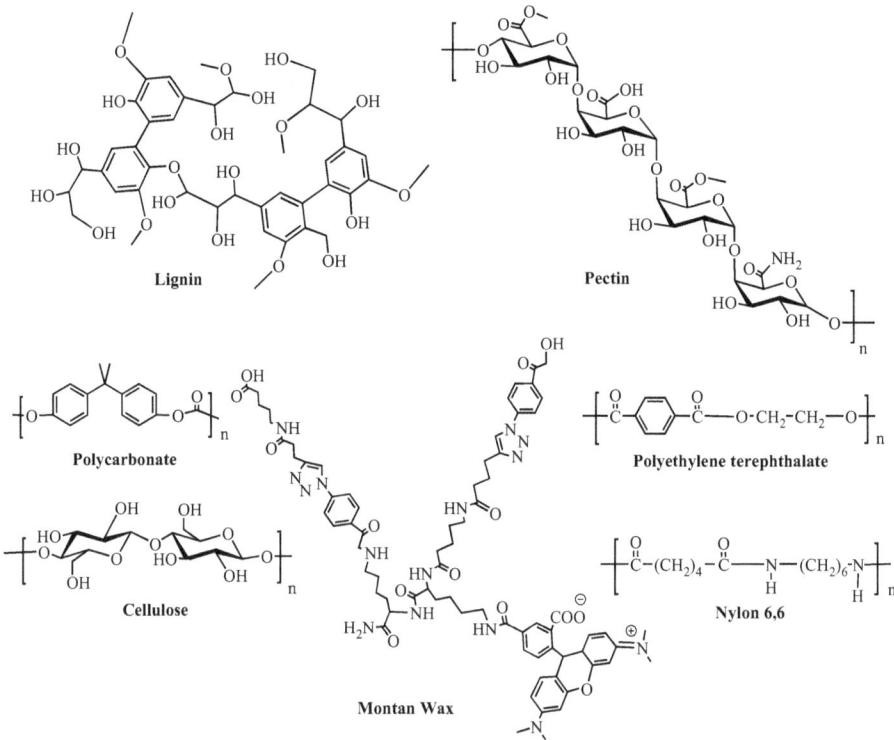

Figure 1.5: Examples of macromolecules and polymers.

based on their origin, structure, polymerization process, and molecular interactions. They are produced through polymerization, a process that converts monomers into polymers. Polymerization is categorized into two types: condensation and addition. The three most common condensation polymers are polyesters, polyamides, and polycarbonates. Addition polymerization is further classified into free-radical polymerization, ionic polymerization (cationic and anionic), and coordination polymerization.

The addition polymerization reaction consists of three steps: initiation, propagation, and termination. There are four polymerization techniques: bulk, solution, suspension, and emulsion.

Numerous synthetic polymers are widely used in the industry. Some examples include polyethylene terephthalate (PET or PETE), polyvinyl chloride (PVC), polylactic acid (PLA), polyamide (PA), polyethylene (PE), polyethylene oxide (PEO), polyacrylic acid (PAA), polypropylene (PP), polypropylene oxide (PPO), polyvinyl acetate (PVA), polycarbonate (PC), and polystyrene (PS). Table 1.4 presents selected examples of natural and synthetic polymers.

Table 1.4: Examples of biopolymers and synthetic polymers, their monomeric units, and the atoms present in their structures.

Polymers	Building blocks	Atoms present in the structure	Types
Polysaccharides	Monosaccharides	C, H, O	**Biopolymers**
Polypeptides	Amino acids	C, H, O, N	
Oligonucleotides	Nucleotides	C, H, O, N, P	
Polylactic acid	Lactic acid	C, H, O	**Synthetic polymers**
Polyvinyl chloride	Vinyl chloride	C, H, Cl	
Polyacrylic acid	Acrylic acid	C, H, O	
Polyvinyl acetate	Vinyl acetate	C, H, O	

1.6 Functional Groups in Polymers

Functional groups are specific groups of atoms or bonds within a molecule that determine its chemical reactivity. They play a crucial role in defining the chemical properties and reactivity of polymers. By carefully selecting the appropriate functional groups during polymerization or by modifying existing polymers with specific groups, manufacturers can tailor materials to meet the exact needs of various industries, including biomedical applications, construction, electronics, and packaging. Understanding these groups and their roles provides valuable insight into polymer chemistry and material science. In polymers, functional groups are typically present in the monomeric units or are introduced during polymerization. These groups can influence a polymer's mechanical, thermal, optical, and chemical properties. Table 1.5 provides a detailed overview of common functional groups in polymers, including their names, structures, and applications.

Table 1.5: Common functional groups in polymers, their names, structures, and applications.

Name	Structure	Applications "Participating in the formation of"
Hydroxyl group	(–OH)	Polyethylene glycol and polyvinyl alcohol
Carboxyl group	(–COOH)	Polyethylene terephthalate and polyacrylic acid
Amino group	(–NH$_2$)	Nylon (polyamide) polyurethane
Isocyanate group	(–N=C=O)	Polyurethane and polyurea
Ester group	(–COO–)	Polyethylene terephthalate and polybutylene terephthalate

Table 1.5 (continued)

Name	Structure	Applications "Participating in the formation of"
Epoxy group	(–C–O–C–)	Epoxy resins
Thiol group	(–SH)	Polydimethylsiloxane (thiol groups can be used in silicone cross-linking)
Aldehyde group	(–CHO)	Polyvinyl aldehyde, which is formed by the oxidation of polyvinyl alcohol
Nitroso group	(–NO)	Nitroso-containing polymers
Carbon-carbon double bond	(C=C)	Polyethylene Polypropylene Polystyrene Polyvinyl chloride

1.7 Polymer Classifications

Polymers are versatile materials that can be classified based on a variety of factors, including their physical properties, origin, structure, polymerization method, and applications. This classification facilitates an understanding of their properties, processing methodologies, and usefulness for a variety of applications. Figure 1.6 depicts a broad categorization of polymers, with Table 1.6 providing examples of each category.

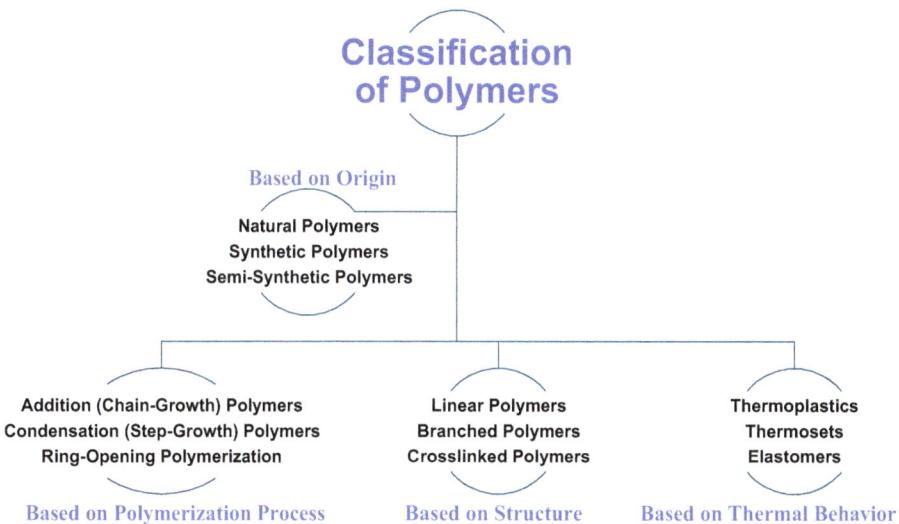

Figure 1.6: General classification of polymers.

Table 1.6: Examples of different polymer classes.

Polymer class	Examples
Natural polymers	Polysaccharides (cellulose and starch) Proteins (collagen and keratin) Nucleic acids (DNA and RNA) Natural rubber
Synthetic polymers	Polyethylene Polystyrene Polyvinyl chloride Nylon (polyamide) Polypropylene
Semi-synthetic polymers	Cellophane Rayon Vulcanized rubber
Addition (chain-growth) polymers	Polyethylene Polystyrene Polyvinyl chloride
Condensation (step-growth) polymers	Polyesters Nylon Polyurethanes
Ring-opening polymers	Polylactic acid Polyethylene glycol
Linear polymers	Polyethylene Polypropylene
Branched polymers	Low-density polyethylene Glycogen Amylopectin
Crosslinked polymers	Vulcanized rubber Epoxy resins
Thermoplastics	Polyethylene Polystyrene Polyvinyl chloride
Thermosets	Epoxy resins Phenolic resins Urea-formaldehyde
Elastomers	Natural rubber Styrene-butadiene rubber Silicone rubber

1.8 Notation and Nomenclature of Polymers

In polymer chemistry, a clear and consistent naming and notation system is essential for accurately understanding and working with polymers. Polymer nomenclature provides a standardized method for naming and representing polymers, facilitating effective communication across scientific and industrial communities. By adhering to standardized conventions, scientists and engineers can efficiently convey information about polymer structures, synthesis, and properties, ensuring clarity and consistency in the field of polymer science.

1.8.1 Monomer Nomenclature

A monomer is the fundamental unit from which a polymer is synthesized. The nomenclature of monomers follows standardized rules established by IUPAC (International Union of Pure and Applied Chemistry) and other regulatory organizations. Monomers are typically named based on their chemical structure and the functional groups they contain. For simple monomers, the name is often derived from the parent compound or its constituent atoms. For example, ethene (C_2H_4) serves as the monomer for polyethylene (PE), while styrene ($C_6H_5CH{=}CH_2$) is used to produce polystyrene (PS). When a monomer contains a functional group (e.g., hydroxyl, carboxyl), the functional group is included in its name. For instance, acrylic acid ($CH_2{=}CH{-}COOH$) is the monomer for polyacrylic acid (PAA).

1.8.2 Polymer Nomenclature

The systematic naming of polymers follows a general rule in which the polymer's name is derived from its monomer, with the prefix "poly-" added, along with a suffix that represents the repeating unit. If the polymer consists of multiple monomers or has a complex structure, its name may reflect these features.

For homopolymers, the name is typically formed by adding the prefix "poly-" to the monomer's name. For example, polyethylene (PE) is derived from ethylene (C_2H_4), and polystyrene (PS) originates from styrene ($C_6H_5CH{=}CH_2$). For copolymers, which are synthesized from two or more different monomers, the names of the monomers are included in the polymer's name. For example, styrene-butadiene rubber (SBR) is a copolymer of styrene and butadiene, while nylon-6,6 is a copolymer of hexamethylenediamine and adipic acid.

1.8.3 Polymerization Notation

The polymerization process is typically represented using a general formula that depicts the repeating unit of the polymer chain. This notation commonly includes square brackets to indicate the repeating unit and a subscript "n" to denote the degree of polymerization, which refers to the number of repeating units in the polymer chain. For instance, $[-A-]_n$ is a fundamental notation for a homopolymer, where "A" signifies the repeating monomer unit, and "n" represents the degree of polymerization. In the case of polyethylene, it is denoted as $[-CH_2-CH_2-]_n$. For copolymers, the two monomers involved are represented in the repeating unit. The structure of the copolymer is denoted by including both monomers in the brackets. For example, a copolymer of styrene and butadiene could be denoted as $[-C_6H_5CH{=}CH_2-]_{x \text{ or } m}[-CH_2CH{=}CH_2-]_{y \text{ or } n}$.

1.8.4 Degree of Polymerization (DP)

The degree of polymerization (DP) refers to the total number of monomer units present in a polymer chain and is often used to estimate the polymer's molecular weight. It is commonly represented as a subscript next to the polymer notation. Notation for DP: $[-A-]_n$, where n denotes the degree of polymerization. For example, if a polymer chain has a DP of 1,000, polyethylene can be represented as $[-CH_2-CH_2-]_{1000}$.

1.8.5 Acronyms or Abbreviations

Polymers are commonly identified by acronyms or abbreviations in both industry and scientific literature. These abbreviations serve as a convenient shorthand, facilitating clear and efficient communication across various fields, including research, production, and application development. They are often derived from the polymer's chemical structure, classification, or synthesis method.

From widely used plastics to specialized high-performance materials, these abbreviations help streamline discussions in technical documentation and industry practices. Below is a short list of commonly used polymer abbreviations, along with their meanings, providing a quick reference for professionals working with polymers.

- Polyacrylic acid (PAA)
- Acrylonitrile butadiene styrene (ABS)
- Polyamide (PA)
- Polybutylene terephthalate (PBT)
- Polycarbonate (PC)
- Polyetheretherketone (PEEK)
- Polyethylene (PE)

- High-density polyethylene (HDPE)
- Low-density polyethylene (LDPE)
- Polyethylene terephthalate (PET)
- Polyethylene glycol (PEG)
- Polyhydroxyalkanoates (PHA)
- Polyimide (PI)
- Polylactic acid (PLA)
- Polymethyl methacrylate (PMMA)
- Polydimethylsiloxane (PDMS)
- Polyoxymethylene (POM)
- Polyphenylene oxide (PPO)
- Polyphenylene sulfide (PPS)
- Polypropylene (PP)
- Polystyrene (PS)
- Polytetrafluoroethylene (PTFE), Teflon
- Polyurethane (PUR)
- Polyurea (PUA)
- Polyvinyl acetate (PVAc)
- Polyvinyl alcohol (PVA)
- Polyvinyl chloride (PVC)
- Polyvinylidene fluoride (PVDF)
- Styrene-butadiene rubber (SBR)

1.8.6 Simplified Structures of Common Bio- and Synthetic Polymers

Biopolymers

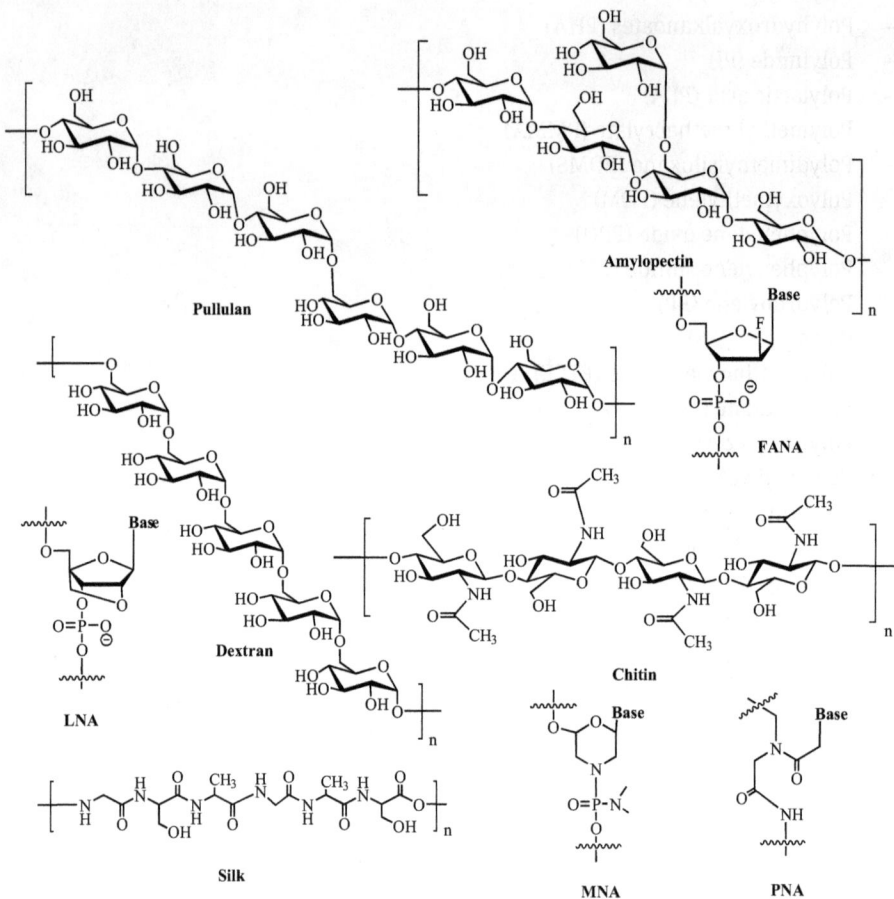

Pullulan

Amylopectin

FANA

Dextran

Chitin

LNA

Silk

MNA

PNA

Synthetic polymers

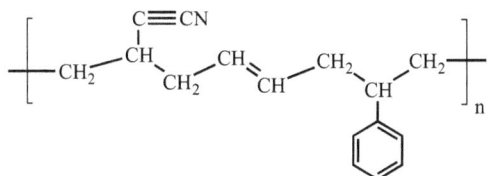

Acrylonitrile-butadiene-styrene (ABS)

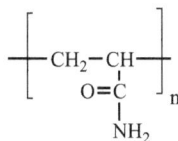

Polyacrylamide (PAM)

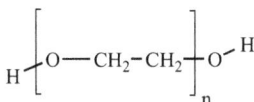

Poly(ethylene glycol) (PEG)

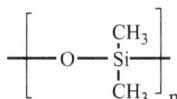

Polydimethylsiloxane (PDMS)

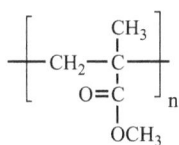

Poly(methylmethacrylate) (PMMA)

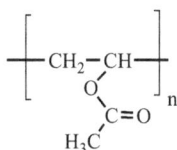

Poly(vinyl acetate) (PVAc)

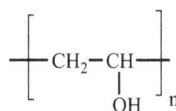

Poly(vinyl alcohol) (PVA)

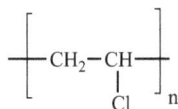

Poly(vinyl chloride) (PVC)

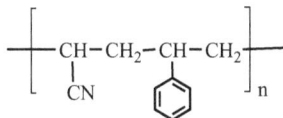

Styrene-Acrylonitrile Resin(SAN)

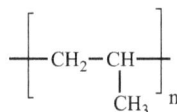

Polypropylene (PP)

1.9 Essential Keywords

Biodegradable polymers Materials that break down naturally through microbial activity.

Biomacromolecules Large, naturally occurring polymers, such as carbohydrates (sugars), proteins (peptides and polypeptides), and nucleic acids (DNA and RNA).

Branched polymers Structures with side chains extending from the main molecular backbone.

Cross-linked polymers Structures in which molecular chains are interconnected by strong covalent bonds, enhancing stability and rigidity.

Degree of polymerization (DP) The total number of monomer units in a polymer chain is often used to estimate molecular weight.

Elastomers Flexible polymers with elastic properties, allowing them to return to their original shape after being stretched.

Engineering plastics A class of synthetic polymers designed for enhanced mechanical properties, such as strength, durability, heat resistance, chemical resistance, and dimensional stability.

Functional groups Specific groups of atoms or bonds within a molecule that determine its chemical reactivity and properties.

Linear polymers Molecules composed of monomer units linked in a straight, unbranched chain.

Macromolecules Large molecules, either polymeric or non-polymeric, with high molecular mass.

Monomers (repeating units) The fundamental building blocks of polymers, which chemically bond through polymerization to form larger polymer structures.

Natural polymers Materials that occur naturally and are typically biodegradable, including substances such as DNA, proteins, starch, and natural rubber.

Polymer A material composed of large molecules (macromolecules) consisting of repeating subunits or building blocks derived from one or more monomer species.

Polymer boom The rapid growth and development of the polymer industry in the twentieth century, especially after World War II, marked a period when polymers evolved from experimental materials to essential industrial components.

Polymerization The chemical process in which monomers bond together to form a polymer.

Polymerization notation A symbolic representation of the polymerization process, typically using a general formula to depict the repeating unit of the polymer chain.

Polymer recycling The process of collecting, processing, and reusing polymer waste to create new products.

Synthetic polymers Man-made polymers produced from petrochemical sources through chemical synthesis. Examples include plastics, nylon, and synthetic rubber.

Styrofoam A foam version of polystyrene, commonly used for insulation and packaging.

Thermosets Polymers that undergo an irreversible curing process when heated and cannot be re-melted or reshaped once set.

Thermoplastics Polymers that soften when heated, allowing them to be reshaped multiple times.

1.10 Questions and Answers

Questions	Answers
1. What are polymers?	Polymers are large macromolecules consisting of repeating structural units, called monomers, which are chemically bonded to form long chains.
2. What is the origin of the word "polymer"?	The word "polymer" is derived from the Greek words *poly* (meaning "many") and *meros* (meaning "part" or "unit"), emphasizing that a polymer is composed of multiple repeating subunits.
3. How are polymers generally classified?	Natural polymers occur naturally in nature, such as cellulose, starch, proteins, and DNA. Synthetic polymers are man-made and include plastics such as polyethylene, polypropylene, and nylon.
4. What are the cellulose derivatives?	Cellulose acetate, cellulose nitrate, methylcellulose, and carboxymethylcellulose.
5. What is Styrofoam?	A foam version of polystyrene.
6. What does ABS stand for?	Acrylonitrile butadiene styrene.
7. What is a polymer boom?	The rapid expansion and development of the polymer industry during the twentieth century, particularly after World War II, marked the transition of polymers from experimental materials to essential components in various industries.
8. Give examples of synthetic polymers that were discovered during the polymer boom.	Polyethylene (PE), polypropylene (PP), high-density polyethylene (HDPE), acrylonitrile butadiene styrene (ABS), polystyrene (PS), and polyvinyl chloride (PVC).
9. Define the engineering plastics.	A class of synthetic polymers is engineered to exhibit enhanced mechanical properties, including strength, durability, heat resistance, chemical resistance, and dimensional stability.
10. What polymer is considered the precursor to modern engineering plastics?	Bakelite.
11. What are monomers, and how do they form polymers?	Monomers are small molecules that are chemically linked together through the polymerization process to form polymers.
12. List some of the areas where engineering plastics have made a significant impact.	Automotive industry, aerospace, medical devices, and electronics.
13. What was the first recognized example of a polymer blend?	A blend of natural rubber and synthetic rubber was created to address the shortage of natural rubber.

(continued)

Questions	Answers
14. How do polymers affect the environment?	Polymers, particularly plastics, play a major role in environmental pollution. Non-biodegradable plastics persist in landfills, oceans, and ecosystems for centuries, leading to long-term environmental damage.
15. What is the importance of biodegradable polymers?	Biodegradable polymers are crucial because they decompose more easily in the environment than traditional plastics, helping to reduce long-term pollution and minimize environmental impact.

Chapter 2
Chemistry of Polymers

2.1 Polymerization Chemistry

2.1.1 Addition Polymerization

Addition polymerization, or chain-growth polymerization, is one of the most common methods used to synthesize polymers. In this type of polymerization, the monomer contains a double bond (typically carbon-carbon, C=C), which can undergo reactions to form a polymer chain. This process is characterized by the successive addition of monomer molecules to a growing polymer chain, without the loss of any atoms or molecules during the reaction, and the repeating unit of the polymer contains the same atoms as the monomer. The majority of addition polymers, as shown in Table 2.1, are formed from monomers that have at least one double C=C bond, as illustrated in Figure 2.1.

Table 2.1: Examples of addition polymers and their preferred methods of polymerization.

Monomer	Structure	Polymer	Structure	Polymerization methods
Acrylonitrile	$CH=CH_2$ / CN	Polyacrylonitrile	$\left[CH-CH_2\right]_n$ / CN	– Free radical – Anionic
1,3-Butadiene	CH_2 / CH_2 / $C-C$ / H / H	Polybutadiene	$\left[CH_2-CH=CH-CH_2\right]_n$	– Free radical – Anionic
Ethylene	$CH_2=CH_2$	Polyethylene	$\left[CH-CH_2\right]_n$ / H	– Free radical – Cationic – Coordination
Isobutylene	CH_3 / $C=CH_2$ / CH_3	Polyisobutylene	$\left[C-CH_2\right]_n$ / CH_3 / CH_3	– Cationic
Isoprene	CH_2 / CH_2 / $C-C$ / CH_3 / H	Polyisoprene	$\left[CH_2-C=CH-CH_2\right]_n$ / CH_3	– Free radical – Anionic – Coordination
Methyl methacrylate	CH_3 / $C=CH_2$ / $COOCH_3$	Polymethyl methacrylate	$\left[C-CH_2\right]_n$ / CH_3 / $COOCH_3$	– Free radical – Anionic – Coordination
Propylene	$CH=CH_2$ / CH_3	Polypropylene	$\left[CH-CH_2\right]_n$ / CH_3	– Cationic – Coordination

https://doi.org/10.1515/9783111585734-002

Table 2.1 (continued)

Monomer	Structure	Polymer	Structure	Polymerization methods
Styrene	$CH=CH_2$ C_6H_5	Polystyrene	$\left(-CH-CH_2-\right)_n$ C_6H_5	– Free radical – Anionic – Cationic – Coordination

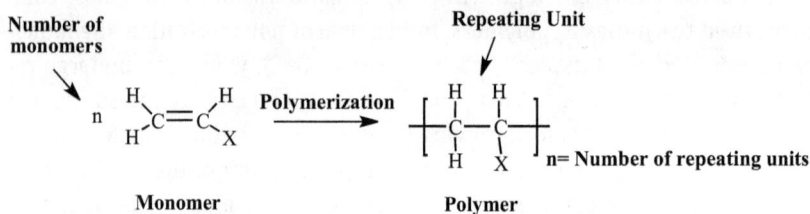

Figure 2.1: Addition polymerization general scheme.

The general process of addition polymerization involves three distinct stages:

Initiation: In this step, active species such as a free radical, cation, or anion are required to start the polymerization process (Figure 2.2). These active species are commonly generated by initiators depending on the nature of initiation. Examples of free radical initiators are benzoyl peroxide (BPO) and azobisisobutyronitrile (AIBN). Cationic initiators include Lewis acids such as boron trifluoride (BF_3), aluminum chloride ($AlCl_3$), titanium tetrachloride ($TiCl_4$), and tin tetrachloride or stannic chloride ($SnCl_4$). Anionic initiators include alkoxides, hydroxides, cyanides, phosphines, amines, and organometallic compounds such as n-butyl lithium (n-butyl Li) and sec-butyl lithium (sec-butyl Li). Initiators for coordination polymerization are usually heterogeneous Ziegler-Natta catalysts based on titanium tetrachloride and organoaluminum co-catalysts.

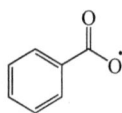

Free radical   Carbanion   Carbocation

Figure 2.2: Types of active species in addition polymerization.

The free radical initiator is a compound that can produce free radicals. These radicals are highly reactive species with an unpaired electron (an odd number of electrons), and they can initiate the polymerization reaction by breaking the monomer's double bond. The free radical initiator attacks the double bond in a monomer, breaking the C=C bond and creating a new free radical at the end of the monomer. The cationic initiator transfers charge to a monomer, which then becomes reactive and forms a polymer. On the other hand, an anionic initiator transfers an electron (or a negatively charged group) to a monomer, which then becomes reactive.

Propagation: After the initial monomer is introduced, the growing polymer chain continues to incorporate additional monomers during the propagation stage. In this process, the polymer chain extends as the double bond of each incoming monomer is repeatedly broken. The free radical (or cation/anion) at the chain's end reacts with another monomer, breaking its C=C bond and integrating it into the chain. The newly added monomer forms a new active site at the end of the expanding polymer chain, enabling further monomer addition. This cycle repeats multiple times until the polymer reaches the desired length and molecular weight.

Termination: This is the final step, occurring through combination, disproportionation, or chain transfer reactions.

There are various types of addition polymerization, each with numerous applications. Table 2.2 provides a brief summary of these types and their respective applications.

Table 2.2: Types and applications of addition polymerization.

Types
Free radical polymerization: The most common type of addition polymerization uses a free radical initiator to start the process. Examples of polymers made by this type include polyvinyl acetate (PVAc), polystyrene (PS), and poly(methyl methacrylate) (PMMA).
Cationic polymerization: Initiated by a positively charged species (cation), it is often used for monomers with electron-withdrawing groups. Examples of polymers made by this type include polystyrene (PS), polypropylene (PP), and polyisobutylene (PIB).
Anionic polymerization: Initiated by a negatively charged species (anion), it is typically used for monomers with electron-donating groups. Examples of polymers made by this type include styrene-butadiene rubber (SBR) and neoprene, also known as polychloroprene.
Coordination (Ziegler-Natta) polymerization: A special type of addition polymerization that uses metal catalysts (such as Ziegler-Natta catalysts) to control polymerization. It is commonly used for olefins (e.g., ethylene, propylene). Examples of polymers made by this type include high-density polyethylene (HDPE) and isotactic polypropylene (iPP).

Table 2.2 (continued)

Applications

Plastics and packaging: Polyethylene (PE), polypropylene (PP), and polystyrene (PS) are common addition polymers used in packaging materials, containers, and consumer goods.
Synthetic rubbers: Polymers such as polybutadiene and styrene-butadiene rubber (SBR) are used in tires, footwear, and other rubber-based products.
Medical devices: Polymers such as polymethyl methacrylate (PMMA) are used in medical devices, including contact lenses and bone cement.
Coatings and adhesives: Polymers like polyvinyl chloride (PVC) are often used for coatings, and styrene-butadiene rubber (SBR) is used as adhesives.

2.1.1.1 Free Radical Polymerization

Free radical polymerization (FRP) is a widely used technique for the production of a vast range of polymers. It is characterized by its simplicity, versatility, and ability to produce high-molecular-weight materials. The mechanism of free radical polymerization, as shown in Figure 2.3, involves the generation of free radicals, followed by propagation, where the polymer chain grows by the successive addition of monomer units. This process continues until termination occurs, and the chain stops growing. Termination can take place by combination, disproportionation or chain transfer. In the chain transfer step of polymerization termination, the activity of the growing polymer is transferred to another monomer, another polymer, a solvent, or a chain transfer agent with at least one weak chemical bond that can be broken to facilitate the chain transfer reaction. Examples of chain transfer agents include n-octyl mercaptan (NOM), n-dodecyl mercaptan (NDM or DDM), 1,8-dimercapto-3,6-dioxaoctane (DMDO), and carbon tetrachloride (CCl_4).

Although free radical polymerization is simple and highly adaptable to a wide range of monomers, making it suitable for the production of a diverse range of polymers, allows for the synthesis of high-molecular-weight polymers, which can be critical for material properties such as strength, durability, and processing. It can also be used to polymerize a variety of monomers, including vinyl monomers, acrylics, and styrenes, producing a broad spectrum of materials. However, the molecular weight distribution (polydispersity) of the polymer can be wide, especially when the polymerization is not carefully controlled. This can result in less predictable material properties, and the occurrence of termination events (combination or disproportionation) can lead to the formation of different chain lengths, thereby affecting the overall polymer properties.

2.1.1.2 Cationic Polymerization

Cationic polymerization is a type of chain-growth polymerization in which initiators produce cationic species (positively charged ions) that combine with monomers to form

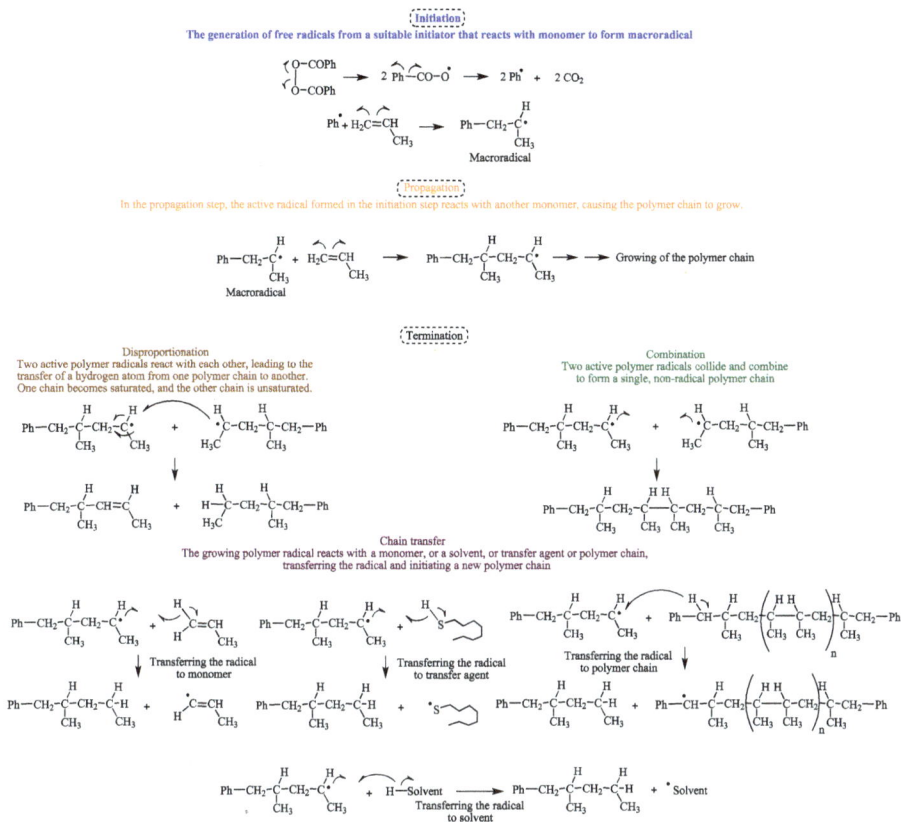

Figure 2.3: Free-radical polymerization mechanism.

a developing polymer chain. Traditional initiators (polymerization catalysts) include Lewis acids such as boron trifluoride (BF_3), tin tetrachloride ($SnCl_4$), and aluminum chloride ($AlCl_3$), as well as acids like sulfuric acid (H_2SO_4), phosphoric acid (H_3PO_4), and trichloroacetic acid (CCl_3COOH). This polymerization process is widely utilized for vinyl monomers, particularly those with electron-withdrawing groups like styrene, isobutene, and butadiene. Cationic polymerization is a highly effective method for manufacturing specific polymers, offering strong stereoselectivity, fast polymerization rates, and fine control over molecular weight and polymer design.

While cationic polymerization offers several advantages, such as control over molecular weight and polymer structure, it also presents challenges, such as impurity sensitivity, monomer limitations, and broad molecular weight distributions if not carefully controlled. The cationic polymerization mechanism involves three main stages: initiation, propagation, and termination. These steps are driven by the presence of cationic species (positively charged intermediates), as shown in Figure 2.4, which facilitate the polymerization process.

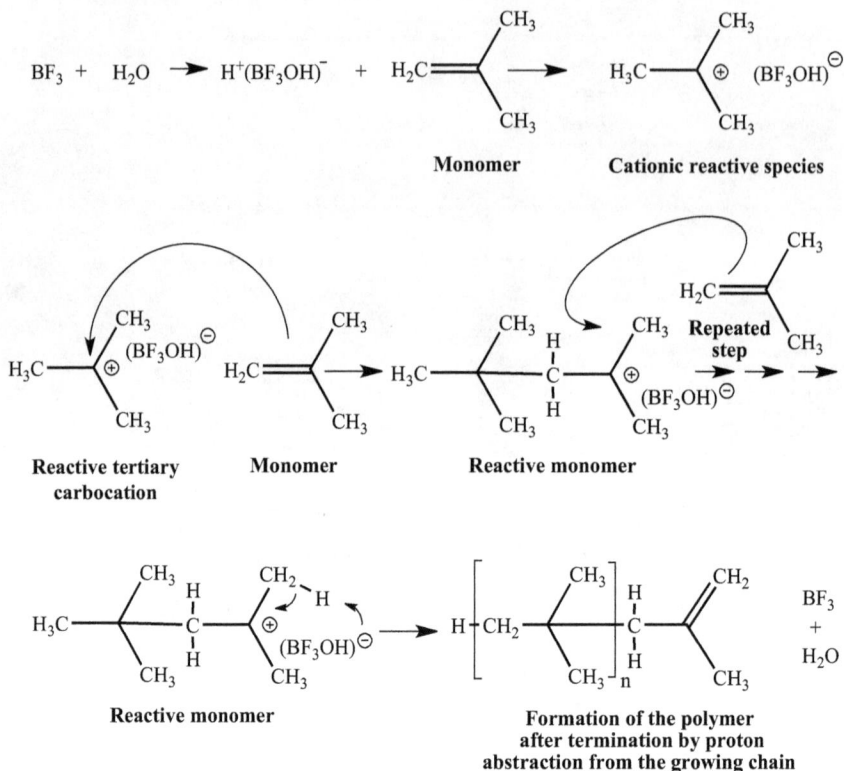

Figure 2.4: Cationic polymerization.

Cationic polymerization has many advantages. It allows for the production of polymers with controlled molecular weights, high stereoselectivity, and specific tacticity (isotactic or syndiotactic configurations). On the other hand, cationic polymerization has some limitations. Among these limitations, not all monomers can be polymerized via cationic polymerization. The process tends to favor electron-rich monomers with electron-withdrawing groups, and the polymerization of some monomers can lead to uncontrolled reactions. Water, oxygen, and other impurities can also inhibit or interfere with cationic polymerization by reacting with the cationic species and stopping the polymerization. This requires a controlled environment, often under dry conditions and in an inert atmosphere. Additionally, it tends to be thermally sensitive and often requires low temperatures to prevent side reactions that might interfere with polymerization.

2.1.1.3 Anionic Polymerization
Anionic polymerization is a type of chain-growth polymerization in which the polymerization process is initiated by the generation of an anionic species (a negatively charged

ion), as shown in Figure 2.5. This negatively charged species, typically formed by a strong base, initiates polymerization by attacking a monomer that can stabilize the negative charge at the polymer's growing end. Anionic polymerization monomers commonly contain electron-withdrawing groups such as –CN and –COOH. Anionic polymerization is a powerful tool for the production of highly controlled, living polymers. It provides precise control over molecular weight, narrow polydispersity, and the ability to synthesize functionalized polymers with specific architectures, such as block copolymers. In anionic polymerization, soluble ionic initiators at low temperatures in polar solvents tend to produce syndiotactic polymers, while in non-polar solvents they produce isotactic polymers. The termination step does not occur spontaneously; it can be achieved by adding ammonia or unintentionally by traces of impurities.

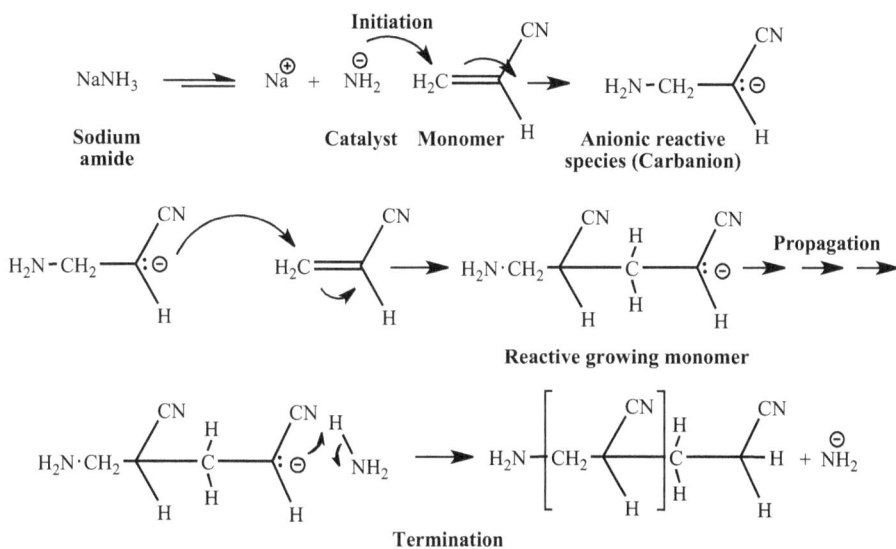

Figure 2.5: Anionic polymerization.

While there are challenges, particularly with monomer limitations and sensitivity to moisture, anionic polymerization remains a cornerstone in the synthesis of specialty polymers for applications in plastics, rubbers, coatings, and medical technologies. It also has many advantages. One of the most significant advantages of anionic polymerization is its ability to produce living polymers – polymers that can be continuously grown, allowing for controlled molecular weight and low polydispersity. This makes it ideal for synthesizing block copolymers. Anionic polymerization can produce functionalized polymers with controlled end groups, enabling further chemical modification or crosslinking in later stages. It also exhibits high reactivity, allowing for rapid polymerization of active monomers and enabling the production of highly reactive functional polymers. Despite its many advantages, anionic polymerization also has limitations. Among these limitations is its sensitivity to moisture and oxygen, as these can protonate the anionic species and terminate the polymerization. As a result, reactions must be conducted under dry, inert conditions (e.g., a nitrogen or argon atmosphere). While anionic polymerization is fast, it requires precise control of the reaction environment and reagents, which can pose challenges for large-scale industrial production.

2.1.1.4 Coordination Polymerization

Ziegler-Natta polymerization is one of the most widely used methods for the polymerization of olefins, such as ethylene and propylene. It involves the use of a metal complex catalyst, often based on titanium tetrachloride ($TiCl_4$), combined with an organometallic co-catalyst, such as triethylaluminum ($AlEt_3$) or diethylaluminum chloride ($AlEt_2Cl$). The Ziegler-Natta catalyst enables the polymerization of α-olefins by controlling the stereochemistry of the resulting polymer and producing high-molecular-weight materials.

Coordination polymerization is the process of coordinating a monomer to a metal center, which results in activation and subsequent polymer chain propagation. This approach is very useful for creating high-quality, high-molecular-weight polymers with precise structure and tacticity. The mechanism of coordination polymerization varies somewhat depending on the monomer and catalyst used. During the initiation stage, an organoaluminum molecule interacts with $TiCl_4$ to create a highly reactive titanium alkyl complex. This active complex then coordinates with an olefin monomer, breaking the π-bond and providing a reactive site for further polymerization.

During propagation, the metal center (typically Ti) forms a coordinate bond with the monomer, and the monomer inserts into the metal-carbon bond. This results in the elongation of the polymer chain. The chain grows one monomer at a time by the coordination of each new monomer with the metal center. The polymerization continues as the metal center keeps reacting with incoming monomers. Termination can be achieved through three approaches. The first is by β-elimination from the polymer chain, forming a metal hydride. The second is by β-elimination with hydrogen transfer to the monomer, and the third is by hydrogenation. Figure 2.6 briefly describes all the steps involved in the Ziegler-Natta coordination polymerization mechanism.

The Ziegler-Natta process of coordination polymerization is capable of producing isotactic, syndiotactic, or atactic polymers, depending on the stereochemistry of the metal-ligand complex. The stereochemistry influences the crystallinity and thermal properties of the polymer. There are many advantages to coordination polymerization.

Among these is the control of stereochemistry (isotactic, syndiotactic, atactic), which influences the polymer's physical properties, including crystallinity, thermal stability, and mechanical properties. This type of polymerization enables the production of high-molecular-weight polymers, which is important for applications requiring strong, durable materials. Metal-based catalysts used in coordination polymerization can be highly

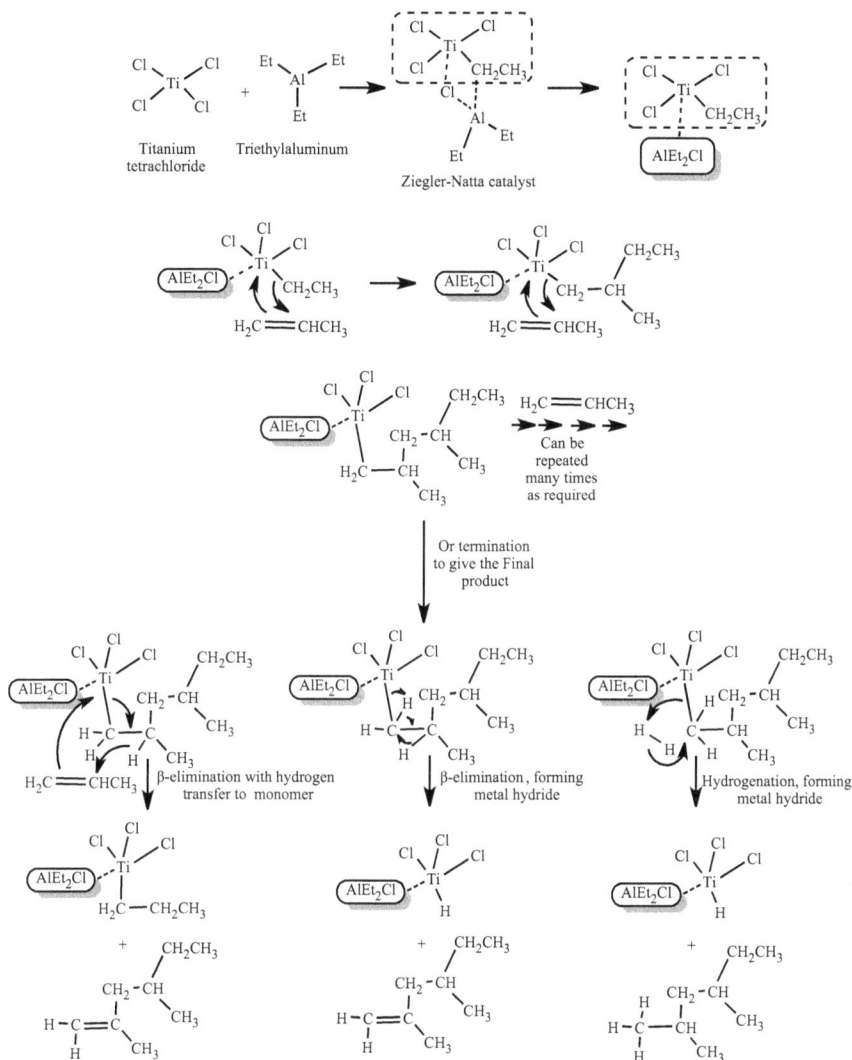

Figure 2.6: Ziegler-Natta coordination polymerization.

efficient and can tolerate a variety of monomers, leading to efficient and cost-effective polymerization processes. Ziegler-Natta polymerization is extensively used to produce polyolefins, including polyethylene (PE), polypropylene (PP), and other olefins with controlled molecular weight and stereochemistry. Examples include high-density polyethylene (HDPE), linear low-density polyethylene (LLDPE), and ultra-high molecular weight polyethylene (UHMWPE).

2.1.2 Ring-Opening Polymerization

Ring-opening polymerization (ROP) is a type of polymerization in which a cyclic monomer (a monomer with a ring structure) undergoes a reaction to open the ring and form a linear polymer. This type of polymerization is widely used for creating macromolecules with precise control over molecular weight and structure. The ring-opening process is distinct from both addition and condensation polymerizations in that it involves the breaking of a ring structure, usually under the influence of a catalyst or other reactive species. ROP generally proceeds through several well-defined mechanisms: free radical polymerization, cationic polymerization, anionic polymerization, and coordination polymerization. Each mechanism involves different types of reactive species to initiate and propagate the polymerization process, but the general concept is the same: the cyclic monomer opens to form a chain with an active center, which can then react with additional monomers (see Figure 2.7).

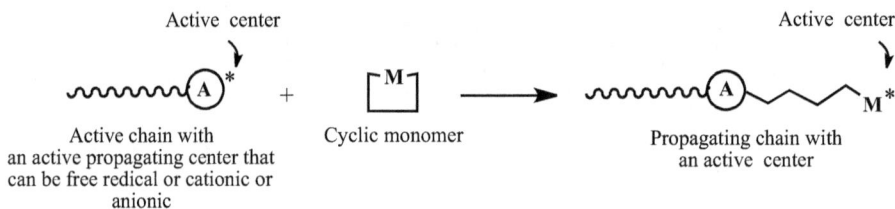

Active center

Active center

+

Active chain with
an active propagating center that
can be free redical or cationic or
anionic

Cyclic monomer

Propagating chain with
an active center

Figure 2.7: General concept of ring-opening polymerization.

In free radical ROP, the polymerization begins with the generation of a free radical that can attack the ring of a cyclic monomer, opening the ring and starting the polymerization. On the other hand, in cationic ROP, the polymerization is initiated by a cationic species, such as a proton (H^+) or an alkyl cation. This cation reacts with a monomer to open the ring and initiate the polymerization. The third type is the anionic ROP, where the polymerization is initiated by an anion, such as an alkoxide (RO^-) or a carboxylate ion.

This anion attacks the electrophilic monomer, breaking its ring and initiating the polymerization. In coordination with ROP, metal catalysts (such as Ziegler-Natta catalysts) are used to initiate the polymerization. These catalysts provide high control over the polymerization and often result in high molecular weight polymers with a narrow distribution of chain lengths.

The ROP can be applied to a wide range of cyclic monomers, each yielding a different type of polymer. Some common types of cyclic monomers used in ROP include cyclic ethers, cyclic esters, cyclic carbonates, cyclic amides, cyclic thionoesters, cyclic sulfur-containing carbonates, cyclic xanthates, and thiiranes. Selected examples are listed in Table 2.3.

Table 2.3: Selected examples of the common types of cyclic monomers used in ROP.

Cyclic monomer	Polymer
Cyclic Ethers	
Ethylene oxide (oxirane)	
Cyclic Esters	
Oxetan-2-one	
Cyclic Carbonates	
1,3-Dioxolan-2-one	
Cyclic Amides	
Pyrrolidin-2-one	
Cyclic Thionoesters	
Dihydrofuran-2(3*H*)-thione	(OR) (Through S-O Isomerization)
Sulfur Containing Carbonates	
1,3-Oxathiolan-2-one	

Table 2.3 (continued)

Cyclic monomer	Polymer
Cyclic Xanthates	
5-Alkyl-1,3-oxathiolane-2-thione	(Through S-O Isomerization)
Thiiranes	
2-Methylthiirane	

ROP is widely used in various industries to produce specialized polymers with unique properties. Some of the key applications include Nylon-6 and Nylon-6,6 production, bio-degradable polymers such as polylactic acid (PLA), polyesters for medical applications such as polycaprolactone (PCL), which is used in tissue engineering and drug delivery systems, and engineering plastics such as polysiloxanes and silicones, which are used in electrical insulators, adhesives, coatings, and medical implants. Silicones, such as polysiloxanes, are utilized in electrical insulators, adhesives, coatings, and medical implants.

There are many advantages to consider for ring-opening polymerization. Among them are the use of specific catalysts or initiators that allow precise control over the molecular weight of the polymer and the distribution of chain lengths, as well as the use of a wide range of cyclic monomers, enabling the synthesis of polymers with diverse chemical structures and properties.

Despite the previously mentioned advantages, ring-opening polymerization (ROP) faces several challenges. These include:

- Monomer Availability: Some cyclic monomers are either expensive or difficult to synthesize in large quantities.
- Strict Reaction Conditions: ROP often requires carefully controlled conditions, including specific temperatures, solvents, and catalysts, which can complicate the process and increase costs.
- Monomer Stability: Certain cyclic monomers, particularly epoxides and lactams, are sensitive to air, moisture, or other contaminants, which can affect the efficiency and outcome of the polymerization.

2.1.3 Condensation Polymerization

Condensation polymerization is a widely used method for synthesizing high-performance polymers such as polyesters, polyamides, and polyurethanes. It is characterized by the stepwise reaction of monomers with two or more functional groups, leading to the release of small molecules like water. The ability to form strong, durable, and versatile polymers makes condensation polymerization an essential process in industries ranging from textiles to packaging and coatings. In condensation polymerization, monomers must have bifunctionalities. Each monomer must have at least two reactive sites (functional groups). Monomers that have only one functional group each cannot be used for condensation polymerization because the reaction cannot proceed further. However, if each reacting molecule has two functional groups, then the reaction can continue. The two bifunctional monomers react together to form what is called a dimer. In this case, the product still contains two functional groups, allowing further reaction with monomers to take place to form a trimer and so on, as illustrated in Figure 2.8.

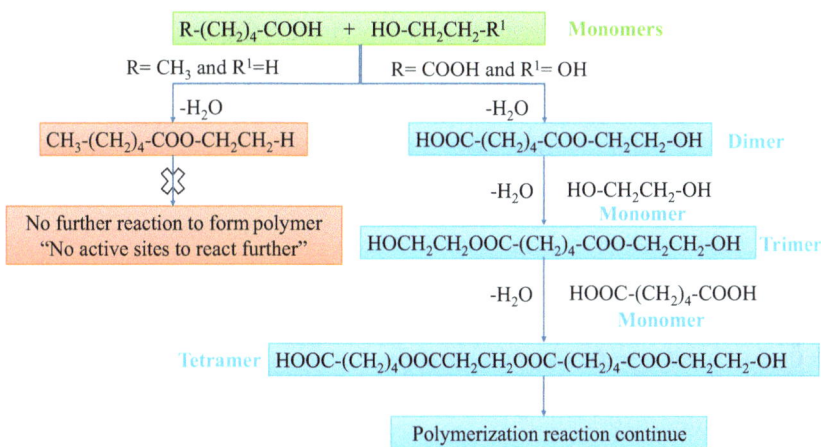

Figure 2.8: General concept of condensation polymerization.

The mechanism of condensation polymerization proceeds through three key stages, similar to addition polymerization, but with a key difference in how monomers react and the elimination of a by-product. In the initiation step, two monomers with two or more reactive functional groups react to form a dimer. This dimerization reaction typically releases a small molecule, such as water, alcohol, or hydrogen chloride.

In the propagation step, the newly formed dimer reacts with another molecule of a monomer (or dimer) containing a functional group. This reaction continues to grow the polymer chain by adding one unit at a time, eliminating a small molecule at each step. As the polymer chain grows, the number of repeating units increases, and the polymer becomes a long-chain molecule with the repeating units linked by covalent bonds. The key feature here is that every step in the polymerization process releases a small molecule, such as water. The termination of condensation polymerization can occur in several ways, such as when a monomer or dimer with a reactive functional group is exhausted, or when the polymer chain stops growing. This could happen for a variety of reasons, such as the functional group becoming blocked, a decrease in the reactivity of the monomer, or the inability to find another reacting partner.

Condensation polymerization is used to produce a wide variety of polymers, including polyesters, polyamides, polycarbonates, and polyurethanes. Each of these polymers is formed through reactions between specific functional groups. These types are discussed in more detail in the following paragraphs.

2.1.3.1 Polyesters

Polyesters are one of the most common types of condensation polymers. The reaction involves the formation of ester linkages between two monomers: typically, a diol (with two hydroxyl groups) and a dicarboxylic acid (with two carboxyl groups). An example is the polymerization of ethylene glycol ($HO-CH_2CH_2-CH2-OH$) and terephthalic acid ($HOOC-C_6H_4-COOH$) to form polyethylene terephthalate (PET) (Figure 2.9).

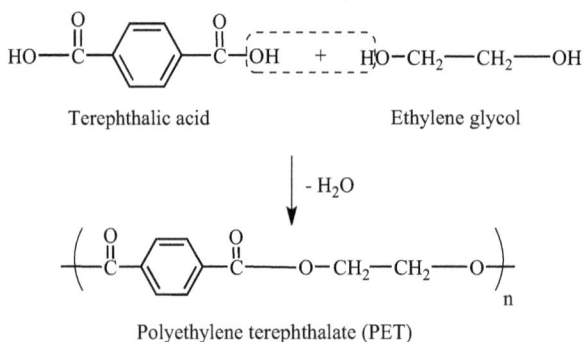

Figure 2.9: Formation of polyethylene terephthalate polymer through condensation polymerization.

2.1.3.2 Polyamides (Nylons)

Polyamides, commonly known as nylons, are produced by the reaction between diamines (compounds with two amino groups, $-NH_2$) and dicarboxylic acids. The reaction between these two monomers forms an amide bond, with the elimination of water. An example is the polymerization of hexamethylenediamine ($H_2N-(CH_2)_6-NH_2$) and adipic acid ($HOOC-(CH_2)_4-COOH$), which forms nylon-6,6 (Figure 2.10).

$$HO-\overset{\overset{O}{\|}}{C}-(CH_2)_4-\overset{\overset{O}{\|}}{C}-OH \quad + \quad H_2N-(CH_2)_6-NH_2$$

Adipic acid Hexamethylenediamine

$\downarrow$ - H_2O

$$\left(\overset{\overset{O}{\|}}{C}-(CH_2)_4-\overset{\overset{O}{\|}}{C}-\overset{}{\underset{H}{N}}-(CH_2)_6-\overset{}{\underset{H}{N}}\right)_n$$

Nylon 6,6

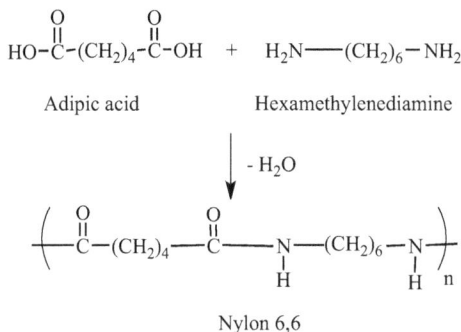

Figure 2.10: Formation of nylon 6,6 polymer through condensation polymerization.

2.1.3.3 Polyurethanes

Polyurethanes are produced by the reaction of diisocyanates (compounds with two iso-cyanate groups, –NCO) with diols. The reaction between the isocyanate and hydroxyl groups results in the formation of a urethane bond (also called a carbamate bond), with the release of carbon dioxide. An example is the polymerization of methylene diphenyl diisocyanate (MDI) with ethylene glycol, which forms a polyurethane (Figure 2.11).

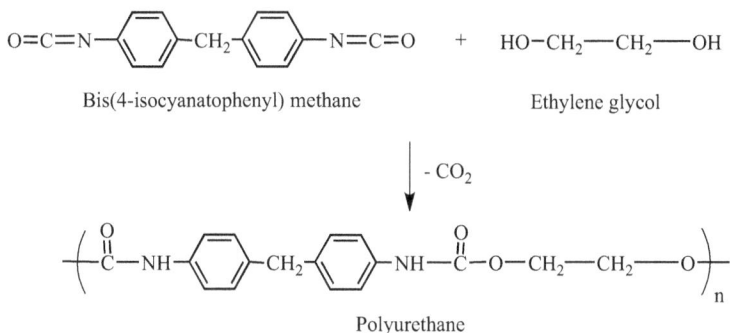

$$O=C=N-\langle\bigcirc\rangle-CH_2-\langle\bigcirc\rangle-N=C=O \quad + \quad HO-CH_2-CH_2-OH$$

Bis(4-isocyanatophenyl) methane Ethylene glycol

$\downarrow$ - CO_2

$$\left(\overset{\overset{O}{\|}}{C}-NH-\langle\bigcirc\rangle-CH_2-\langle\bigcirc\rangle-NH-\overset{\overset{O}{\|}}{C}-O-CH_2-CH_2-O\right)_n$$

Polyurethane

Figure 2.11: Formation of polyurethane through condensation polymerization.

2.1.3.4 Polycarbonates

Polycarbonates are polymers composed of organic functional groups connected by car-bonate linkages. The most commonly used type is a thermoplastic characterized by long molecular chains. Polycarbonate is synthesized through condensation polymerization of bisphenol A (BPA) with either carbonyl chloride (phosgene, $COCl_2$) or diphenyl carbonate $((C_6H_5O)_2CO)$, resulting in the release of hydrogen chloride (HCl) or phenol (C_6H_5OH), re-spectively (Figure 2.12).

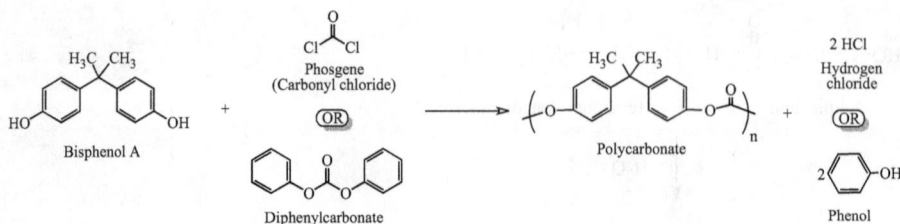

Figure 2.12: Formation of polycarbonate polymer through condensation polymerization.

2.1.4 Copolymerization

Copolymerization is a fundamental technique in polymer chemistry that enables the synthesis of versatile polymers with precisely tailored properties. By strategically selecting monomers and polymerization methods, copolymers can be engineered to fulfill specific requirements across various industrial and consumer applications, including flexible elastomers, adhesives, and high-performance biodegradable materials. The ability to control the structure and characteristics of copolymers is a significant advantage in polymer science, driving innovations in materials design across multiple fields. In the copolymerization process, two or more different types of monomers are polymerized together to form a copolymer. Unlike a homopolymer, which consists of only one type of monomer repeated throughout the chain, a copolymer contains two or more types of monomer units within the polymer chain. The arrangement of these monomer units within the chain can vary, giving rise to different types of copolymers, each with distinct properties. Many commercially important polymers are copolymers. Examples include polyethylene-vinyl acetate (PEVA), nitrile butadiene rubber (NBR), and acrylonitrile butadiene styrene (ABS).

2.1.4.1 Types or Classes of Copolymerization (Copolymers)

There are four main types of copolymerization (or copolymers) based on the arrangement of the different monomers (Figure 2.13) in the copolymer chain. These are:
- Random copolymerization (produces random copolymers).
- Alternating copolymerization (gives alternating copolymers).
- Block copolymerization (produces block copolymers).
- Graft copolymerization (produces graft copolymers).

In **random copolymerization**, the two different monomers are polymerized in a random sequence along the polymer chain. The distribution of the two monomers along the chain is neither regular nor predictable, and their arrangement is governed by the reactivity ratios of the monomers. An example is the copolymer of vinyl chloride and vinyl acetate, made by free-radical copolymerization. Other examples include styrene-butadiene rubber

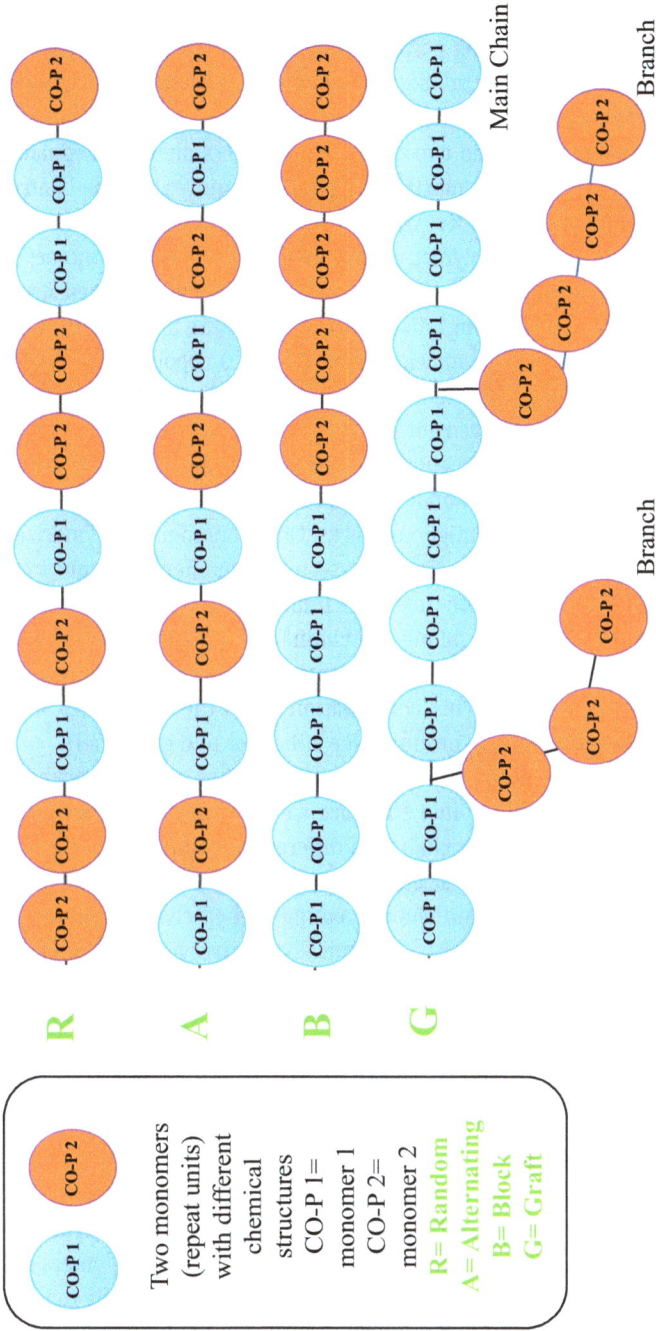

Figure 2.13: Types (classes) of copolymers.

(SBR) and styrene-acrylonitrile resin (SAN). The polymerization involves the simultaneous addition of both monomers to the growing polymer chain, with no specific preference for the order in which the monomers are added. The result is a copolymer in which the monomers are randomly distributed along the chain.

In **alternating copolymerization**, the two different monomers alternate regularly in the polymer chain. This type of copolymerization typically requires precise control over the polymerization conditions and is less common than random copolymerization. One monomer is added to the growing polymer chain, and then the second monomer is added, alternating between them in a regular pattern. This results in a well-defined structure, where each monomer unit appears in a fixed sequence. Styrene and maleic anhydride can copolymerize to form a styrene-maleic anhydride (SMA) copolymer, in which the monomers alternate in the chain, leading to a material that has specific characteristics, such as increased polarity and improved adhesion.

In **block copolymerization**, long segments (or blocks) of one monomer are linked to long segments (or blocks) of another monomer. This results in a copolymer where the monomers are grouped into large, contiguous blocks. One monomer is polymerized first, creating a long block of that monomer. Then, the second monomer is polymerized, creating a second block of the second monomer. The final product consists of two or more "blocks" of monomers that repeat along the chain. Styrene-butadiene-styrene (SBS) is a commonly used block copolymer, where the styrene blocks are typically rigid, while the butadiene block is flexible. This combination gives the copolymer high strength and elasticity, making it useful in products like adhesives and elastomers.

In **graft copolymerization**, one type of monomer is polymerized onto the backbone of another pre-existing polymer, resulting in a structure where branches of one monomer type are attached to the main polymer chain of a different monomer. A main chain polymer (usually a homopolymer) is synthesized first. Then, a second type of monomer is polymerized in the presence of the main chain, creating grafted chains that are attached to the main polymer backbone. This process typically requires a specialized initiator or catalyst to ensure the new monomer is added as branches to the main chain rather than creating a completely new polymer. Polystyrene-graft-poly(methyl methacrylate), PS-g-PMMA, is an example of a graft copolymer.

2.2 Polymerization Kinetics

Polymerization kinetics refers to the study of the rate at which polymerization reactions occur, the factors that influence this rate, and the mechanisms by which polymers are formed. Understanding the kinetics of polymerization is essential for controlling the molecular weight, distribution, and structure of the polymer product. It also helps in optimizing the reaction conditions, improving the efficiency of polymerization processes, and scaling up the production of polymers. Polymerization kinetics can be broken down into several key areas, as shown in Figure 2.14.

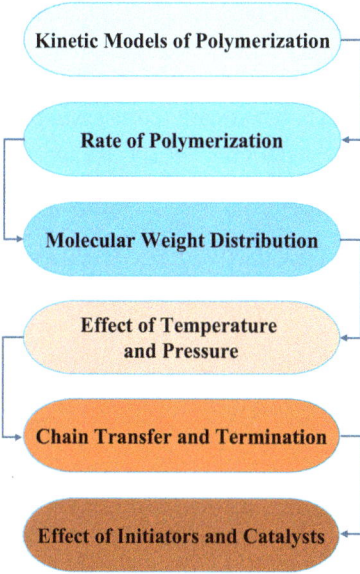

Figure 2.14: Key areas of polymerization kinetics.

2.2.1 Kinetic Models of Polymerization

Polymerization reactions can be categorized into two general types based on the nature of the process: addition (chain-growth) polymerization and condensation (step-growth) polymerization. Each type of polymerization follows different kinetic models. The overall rate of addition polymerization can be expressed using the steady-state approximation, which assumes that the concentration of active polymer chains [P] remains constant throughout the polymerization process:

$$R_P = \frac{k_i[I]^{1/2} k_P[M]}{k_t^{1/2}}$$

where:
R_p is the overall rate of polymerization.
K_i is the rate constant for initiation.
k_p is the rate constant for propagation.
K_t is the rate constant for termination.
[I] is the concentration of the initiator.
[M] is the concentration of the monomer.

Condensation polymerization involves the stepwise reaction between bifunctional or multifunctional monomers, releasing small molecules such as water or methanol during the process. In this case, the rate of polymerization is not governed by a chain mechanism and is generally slower compared to addition polymerization. The kinetics depend on the concentration of both the monomers and the growing polymer chains. The rate of polymerization (R_p) in a step-growth polymerization process is generally first-order with respect to the concentration of each reactive species (monomers) and can be described by:

$$R_p = k[M_1][M_2]$$

where:

R_p is the overall rate of polymerization.
k is the rate constant for the polymerization reaction.
[M_1] and [M_2] are the concentrations of the reacting monomers.

2.2.2 Rate of Polymerization

The rate of polymerization depends on several factors, including the concentration of monomer and initiator, temperature, and the nature of the polymerization system. As the concentration of monomers increases, the rate of polymerization increases because there are more monomer molecules available to react. The concentration of initiators (which create active species) also directly impacts the rate of initiation and propagation. An increase in the initiator concentration typically leads to a higher polymerization rate. The rate of polymerization also increases with temperature. This happens due to the increased kinetic energy of the molecules, which increases the frequency of collisions between monomers and active species. The solvent in which polymerization takes place can also influence the rate of polymerization. Polar solvents may stabilize ionic species, whereas non-polar solvents may be more favorable for free radical polymerization. Catalysts in certain polymerization reactions, such as Ziegler-Natta catalysts for olefin polymerization, can significantly increase the rate of polymerization by lowering the activation energy required for the reaction to proceed.

2.2.3 Molecular Weight Distribution

One of the key aspects of polymerization kinetics is controlling the molecular weight and molecular weight distribution of the polymer. The molecular weight of a polymer is a function of the number of monomer units in the polymer chain and can vary depending on the polymerization conditions. In chain-growth polymerizations, the molecular weight is governed by the rate constants of initiation, propagation, and termination.

The degree of polymerization (DP) is related to the monomer-to-initiator ratio and the termination mechanism. While early termination leads to the formation of short polymers, late termination, in turn, will form long polymers.

In chain-growth polymerizations, the degree of polymerization is the number of monomer units in a polymer chain and is related to the rates of propagation and termination.

$$DP = k_p [M] / k_t$$

The molecular weight (Mw) is related to the degree of polymerization and the molecular weight of the monomer.

$$M_w = DP \times M_0$$

where:

M_0 is the molar mass of the monomer (the molecular weight of the repeating unit).
DP is the degree of polymerization.
M_w is the polymer molecular weight.

In the context of polymerization, molecular weight distribution (MWD) and polydispersity index (PDI) are critical factors in determining the properties and performance of the resulting polymer material. During polymerization, small molecules called monomers are chemically bonded to form long chains known as polymers. However, polymerization doesn't always produce a uniform distribution of chain lengths. The polymer chains vary in length, creating a distribution of molecular weights. This variation arises because polymerization is typically a random process, with some chains growing faster than others or terminating at different times.

MWD describes how polymer chains of varying lengths (and thus molecular weights) are distributed within the polymer sample. In practical terms, this distribution is important because it dictates the material's properties, such as strength, toughness, viscosity, and processing behavior. The MWD refers to the distribution of component polymers that make up a polymer. Examples of narrow and wide molecular weight distributions are shown in Figure 2.15.

The molecular weight distribution is typically described by a few key parameters, such as:

- Number Average Molecular Weight ($\bar{M}_n$): This is the arithmetic mean of the molecular weights of all polymer chains in the sample.

$$\bar{M}_n = \frac{\sum (N_i \cdot M_i)}{\sum N_i}$$

where

N_i is the number of chains with a molecular weight M_i.

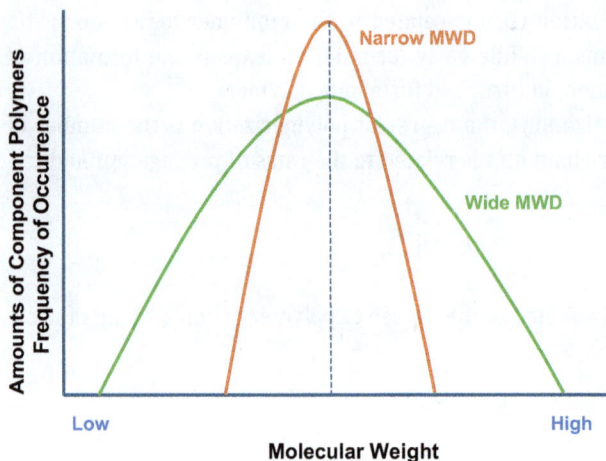

Figure 2.15: Molecular weight distribution (MWD) curve.

– Weight-Average Molecular Weight ($\bar{M}_w$): This is a weighted average that gives more importance to heavier polymer chains.

$$\bar{M}_w = \frac{\sum\left(N_i \cdot M_i^2\right)}{\sum\left(N_i \cdot M_i\right)}$$

The MWD can significantly influence the properties and processing of polymers. For example, a narrow MWD (low PDI) can lead to improved mechanical properties, such as tensile strength and toughness, but may also limit the material's ability to process. A broad MWD (high PDI) may provide better processability and can result in better impact resistance but might compromise certain mechanical properties. In polymer science, it is common to calculate the number average molecular weight ($\bar{M}_n$) and the weight average molecular weight ($\bar{M}_w$) to describe the distribution of polymer chain lengths. The MWD (Figure 2.16) is simply the ratio of the weight average molecular weight ($\bar{M}_w$) to the number average molecular weight ($\bar{M}_n$):

$$PDI = \frac{\bar{M}_w}{\bar{M}_n}$$

PDI = 1: A PDI of 1 indicates a narrow molecular weight distribution, meaning all the polymer chains have nearly the same molecular weight.

PDI > 1: A PDI greater than 1 indicates a broader molecular weight distribution, with a range of chain lengths present in the polymer.

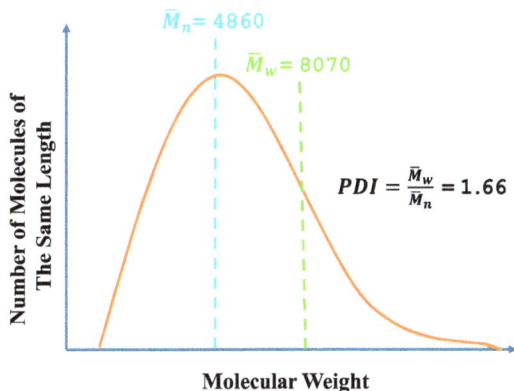

Figure 2.16: Molecular molecular weight distribution (MWD) curve represents the ratio between the weight-average molecular weight ($\bar{M}_w$) and the number-average molecular weight ($\bar{M}_n$).

2.2.4 Effect of Temperature and Pressure

The temperature and pressure during polymerization can significantly affect the polymer's reaction rate and molecular weight. The polymerization rate typically increases as temperature increases, but only up to a certain point. This is because higher temperatures increase the reaction rate by providing more energy to the reactants, thereby increasing the collision frequency and energy between molecules. However, at very high temperatures, undesirable side reactions or decomposition of the polymer may occur. Pressure can also affect the solubility of monomers and influence the polymerization rate, especially in gaseous monomer systems. In the case of bulk polymerization, where monomers are used without a solvent, increasing pressure can help increase the monomer concentration and, thus, the polymerization rate.

2.2.5 Chain Transfer and Termination

In polymerization, chain transfer and termination are crucial steps that affect both the molecular weight and the overall kinetics of the polymerization process. Chain transfer refers to the process in which the active site (usually a radical or ion) of a growing polymer chain is transferred to another molecule. This can lead to the formation of lower molecular weight polymers and impacts the rate of polymerization. Termination occurs when two growing polymer chains combine or when a polymer chain undergoes disproportionation (losing a hydrogen atom to form two different polymer chains). The rate of termination is often influenced by the concentration of active sites in the system.

2.2.6 Effect of Initiators and Catalysts

The choice of initiators and catalysts can dramatically influence the kinetics of the poly-merization process. In free radical polymerization, initiators such as benzoyl peroxide and azobisisobutyronitrile (AIBN) decompose to produce free radicals that initiate the polymerization. The concentration of the initiator determines the number of active sites available for propagation. In catalytic polymerizations, such as Ziegler-Natta polymeriza-tion, catalysts are used to initiate and control the polymerization process by lowering the activation energy, allowing the polymerization to occur under milder conditions.

2.3 Functionalization and Modification of Polymers

2.3.1 Polymer Blending

Polymer blending involves combining two or more distinct polymers to produce a ma-terial that integrates the properties of its components. This technique is extensively uti-lized in the plastics and polymer industry to develop materials with enhanced charac-teristics that surpass those of a single polymer. Polymer blending improves physical properties such as strength, toughness, thermal stability, and chemical resistance, while also optimizing processing behavior. The key reasons for employing polymer blending can be summarized as follows:

- Blending polymers achieves superior characteristics that neither polymer could pro-vide alone. A polymer blend might offer higher impact resistance, better thermal sta-bility, or improved ease of processing.
- A high-performance polymer may be too expensive for a particular application. By blending it with a lower-cost polymer, the overall material cost can be reduced without sacrificing much of the desired performance.
- Blends allow for the customization of material properties, such as altering the ratio of different polymers to achieve the ideal balance of stiffness, flexibility, strength, and chemical resistance.

Polymer blends can be classified into different types based on the interaction be-tween the polymers in the blend. For example, in miscible blends, the two polymers are completely compatible with each other on a molecular level, leading to a single-phase system. The two polymers mix uniformly, and the blend behaves like a single material with a homogeneous composition. On the other hand, in immiscible blends, the two polymers do not mix on a molecular level and instead form distinct phases within the blend (Figure 2.17). Selected examples of miscible and immiscible blends are also listed in Figure 2.18.

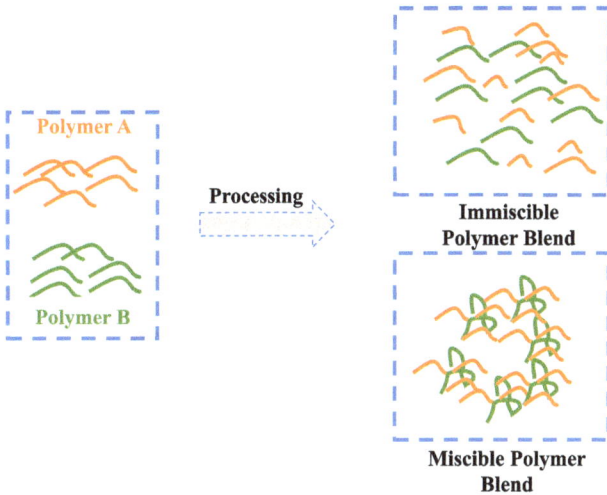

Figure 2.17: General principle for the formation of miscible and immiscible polymer blends.

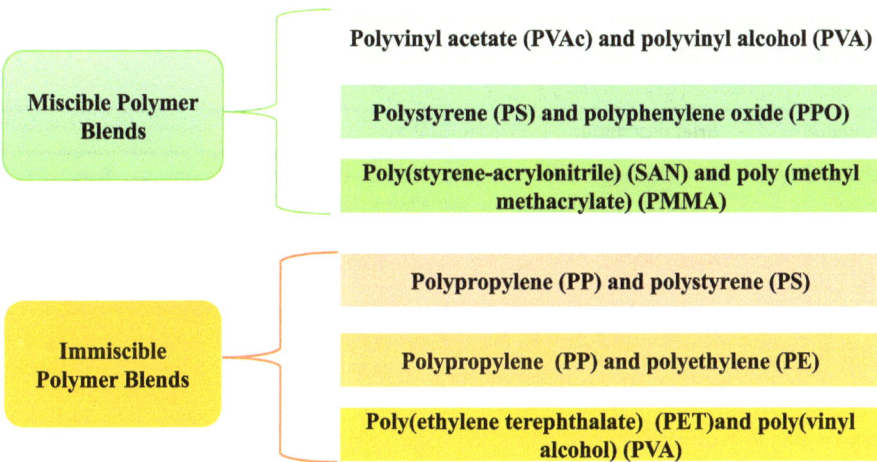

Figure 2.18: Selected examples of miscible and immiscible polymer blends.

2.3.1.1 Polymer Blend Compatibility

The compatibility of polymers in a blend is one of the key factors determining the success of the blending process. Compatibility between the polymers leads to better phase dispersion and improved mechanical properties. Conversely, poor compatibility often results in poor mechanical performance, increased brittleness, and difficulty in processing. Many factors affect polymer compatibility, and among them are:

- Polymers with similar chemical structures, molecular weights, and functional groups are more likely to be compatible.
- The ability of polymers to interact at the molecular level is crucial. Stronger intermolecular forces (e.g., hydrogen bonding, van der Waals forces) often improve compatibility.
- Similar molecular weights between the polymers generally lead to better compatibility because they have comparable viscosities and chain entanglements during processing.
- Polymers with similar glass transition temperature (Tg) values tend to blend more easily, as they have similar thermal behaviors during processing.
- Processing conditions, such as temperature, pressure, and shear forces applied during blending, can influence the degree of mixing and phase separation.

2.3.1.2 Techniques for Polymer Blending

The blending process can be performed using different techniques, depending on the characteristics of the polymers, the targeted properties of the blend, and the production scale. The most commonly used techniques are outlined in Table 2.4.

Table 2.4: Common techniques for polymer blending.

Technique	Brief description
Melt blending or mixing	Melt blending is one of the most widely used methods for polymer blending, especially for immiscible and partially miscible blends. In this process, the polymers are heated until they melt and then mixed using an extruder, a high-shear mixing device that provides heat and pressure to facilitate the blending process.
Solution blending	In solution blending, both polymers are dissolved in a suitable solvent, mixed to form a uniform solution, and then the solvent is evaporated to obtain the polymer blend. This technique is often used when the polymers are not thermally compatible or when they are sensitive to heat.
Latex blending	Latex blending is similar to the solution-blending process; the only difference is that the elastomer is in the form of latex, which presents advantages from an environmental viewpoint.

2.3.2 Polymer Crosslinking

Polymer crosslinking is a chemical process in which individual polymer chains are chemically bonded or linked together to form a network (Figure 2.19). This process alters the physical and mechanical properties of the polymer, often improving its strength, stability, and resistance to deformation. Crosslinking is used to modify the properties of polymers for specific applications, such as in the production of rubber, coatings, adhesives, and

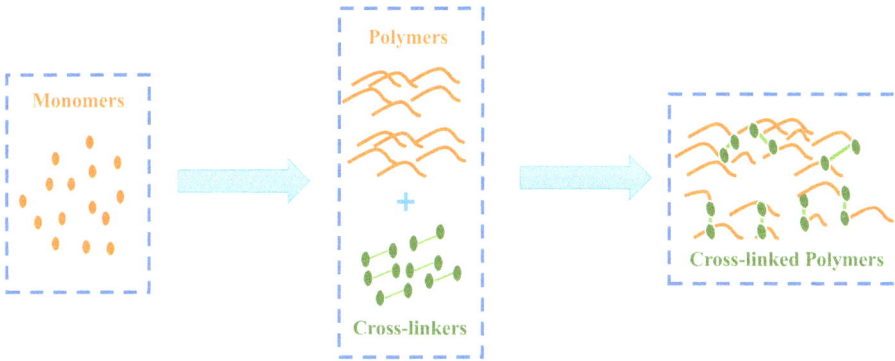

Figure 2.19: General principle of polymer crosslinking.

medical devices. Crosslinking can occur naturally or be induced by external methods such as heat, radiation, or chemical agents. The degree of crosslinking and the method used to achieve it can greatly influence the final properties of the polymer material.

2.3.2.1 Methods of Polymer Crosslinking

Polymer crosslinking can be achieved through various techniques, depending on the polymer type and the desired final properties. Table 2.5 provides a summary of the main crosslinking methods.

Table 2.5: Main crosslinking methods.

Method	Details	Usages
Thermal crosslinking	Heat-induced formation of crosslinks between polymer chains is commonly used with thermosetting polymers such as epoxies, phenolic resins, and rubbers.	Vulcanization of rubber.
Radiation crosslinking	Gamma rays, electron beams, or ultraviolet (UV) radiation induce crosslinking in polymers. Radiation causes the polymer chains to undergo scission (breaking) and recombination, leading to the formation of crosslinks.	Crosslinked polyethylene (PEX).
Chemical crosslinking	Crosslinking induced by chemical agents involves the formation of covalent bonds between polymer chains, either during polymerization or as a post-treatment. These agents react with specific functional groups on the polymer, establishing links that interconnect the chains.	Anion exchange membranes (AEMs) are solid polymer membranes.

Table 2.5 (continued)

Method	Details	Usages
Peroxide crosslinking	Organic peroxide initiator-induced crosslinking occurs when peroxides, such as dicumyl peroxide (DCP), decompose to generate free radicals. These radicals react with polymer chains, promoting the formation of crosslinks.	Crosslinked polyethylene (PEX).
Ionic crosslinking	Ionic crosslinking in acrylic polymers occurs when polymerizable acid groups, derived from carboxylic, phosphonic, or sulfonic acid monomers, interact with multivalent metal ions such as zinc, chromium, titanium, zirconium, and aluminum to form crosslinked structures.	Alginate.
Photo-induced crosslinking	Photo-induced crosslinking (PIC) is an effective method for crosslinking polymers with UV-sensitive functional groups, such as acrylics. This technique is widely used in coatings, adhesives, and printing inks.	Inducing crosslinking among biopolymers and the preparation of edible films.

2.3.2.2 Types of Crosslinked Polymers

Based on the degree of crosslinking, polymers can be categorized into different types, as summarized in Table 2.6.

Table 2.6: Types of crosslinked polymers.

Type	Details	Examples
Crosslinked thermosets	Thermosetting polymers undergo crosslinking upon heating, forming a three-dimensional network that cannot be re-melted or reshaped. Once crosslinked, thermosets are rigid and maintain their shape under high temperatures.	Epoxies, phenolic resins, urethanes, and Bakelite.
Crosslinked thermoplastics	Thermoplastics can also be crosslinked, although they are typically not crosslinked in their standard state. Crosslinking can be induced to enhance the polymer's properties. While thermoplastics can be re-melted, the crosslinked form is more rigid and has higher chemical resistance.	Crosslinked polyethylene (PEX).

Table 2.6 (continued)

Type	Details	Examples
Crosslinked elastomers	Elastomers are highly elastic polymers that can return to their original shape after deformation. Crosslinking is an essential part of their processing to improve their elasticity and shape retention.	Natural rubber or synthetic rubbers, such as butyl rubber or styrene-butadiene rubber (SBR).
Crosslinked hydrogels	Hydrogels are polymers that have been crosslinked to absorb water and form a gel-like structure. Crosslinked hydrogels are used in biomedical applications such as drug delivery systems, wound dressings, and contact lenses.	Polyvinyl alcohol (PVA) and polyethylene glycol (PEG)-based hydrogels.

2.4 Biopolymers and Green Chemistry

The intersection of biopolymers and green chemistry focuses on utilizing sustainable and eco-friendly methods in polymer production and application. Green chemistry seeks to reduce environmental impact across the polymer lifecycle, from raw material sourcing to processing, disposal, and recycling. Meanwhile, biopolymers contribute to sustainability by decreasing dependence on fossil fuels, reducing waste, and improving recyclability and biodegradability.

2.4.1 Biopolymers

Biopolymers are a class of polymers derived from renewable resources, originating from living organisms or biosynthetic processes. These naturally occurring, biodegradable materials often exhibit excellent biocompatibility, making them an eco-friendly alternative to conventional synthetic polymers. Sourced from plants, animals, and microorganisms, biopolymers encompass diverse materials such as proteins, polysaccharides, and nucleic acids. Their sustainability, biodegradability, and reduced environmental impact make them a promising substitute for traditional plastics. However, factors such as high production costs, performance limitations, and supply chain challenges must be addressed to unlock their full potential. With ongoing research and advancements, biopolymers are expected to play an increasingly significant role in various industries, including packaging, agriculture, healthcare, and pharmaceuticals. The three major classes of biopolymers, categorized by their origin and production method, are outlined in Table 2.7.

Table 2.7: Classes of biopolymers and their key applications.

Classes	Examples	Key applications
Natural polymers are obtained directly from biomass.	Polysaccharides (e.g., cellulose, starch, chitosan) and proteins (e.g., collagen, silk, casein).	Packaging, textiles, and biomedical materials.
Polymers that are bio-synthesized using microorganisms and plants, or prepared directly from monomers that are predominantly bio-synthesized.	Polyhydroxyalkanoates (PHA) and polylactic acid (PLA).	Biodegradable plastics, medical implants.
Conventionally, oil-based polymers are prepared from alternative bio-sourced monomers (bioplastics). They are structurally similar to oil-based (petroleum-based) plastics but are mostly not biodegradable and, thus, share the same disposal and recycling issues.	Bio-polyethylene (Bio-PE), Bio-polyethylene terephthalate (Bio-PET), polytrimethylene terephthalate (PTT), and polybutylene succinate (PBS).	Sustainable packaging, automotive parts, and textiles.

Commercially utilized biopolymers, commonly known as bioplastics, are predominantly used in packaging applications. Among them, starch-based plastics and polylactic acid (PLA) are the most widely produced, likely due to their relatively low production costs. In contrast, polyhydroxyalkanoates (PHAs) are manufactured in much smaller quantities due to their high production costs. Table 2.8 presents a selection of commercially used biopolymers, highlighting their applications and biodegradability.

Table 2.8: Commercially used biopolymers, along with their applications and biodegradability.

Biopolymer	Biodegradability (yes or no)	Applications
Starch and blends	Yes	Flexible packaging.
Polyhydroxyalkanoates (PHA)	Yes	Flexible and rigid packaging.
Polybutylene succinate (PBS)	Yes	Flexible packaging.
Polybutylene adipate terephthalate (PBAT)	Yes	Flexible and rigid packaging.
Polylactic acid (PLA)	Yes	Flexible and rigid packaging.
Polyethylene (PE)	No	Flexible and rigid packaging.
Polyethylene terephthalate (PET)	No	Rigid packaging.

There are several misconceptions about biopolymers, one of the most common being that all biopolymers are biodegradable. However, biodegradability is determined by a polymer's chemical structure rather than its production method. While some biopolymers do degrade naturally, others do not, meaning they are not a universal solution to plastic waste accumulation.

Despite claims that biopolymers can solve the plastic crisis, their primary benefit lies in utilizing renewable biomass as a feedstock instead of fossil-based resources like oil and gas. However, bioplastics are mainly suited for packaging applications and cannot fully replace conventional plastics across all industries. Additionally, their large-scale production requires significant agricultural land, potentially affecting food availability for both humans and animals.

2.4.2 Green Chemistry

The green chemistry philosophy is centered on reducing the environmental impact of chemical processes. In the context of biopolymer production, green chemistry principles emphasize the use of renewable feedstocks, minimizing toxic byproducts, reducing energy consumption, and improving the overall sustainability of manufacturing methods. The twelve principles of green chemistry are outlined in Figure 2.20.

Producing biopolymers from renewable raw materials such as plants, algae, or microorganisms reduces dependence on non-renewable fossil fuels and supports a circular economy. For instance, polyhydroxyalkanoates (PHAs) are synthesized by microorganisms that metabolize natural substrates like sugars. This approach also helps eliminate the use of toxic chemicals and solvents common in traditional polymerization, which can generate hazardous waste.

Additionally, biopolymer production processes are optimized for energy efficiency, often requiring lower temperatures and milder reaction conditions than conventional polymer synthesis. Some biopolymers can be synthesized at ambient temperatures, re-

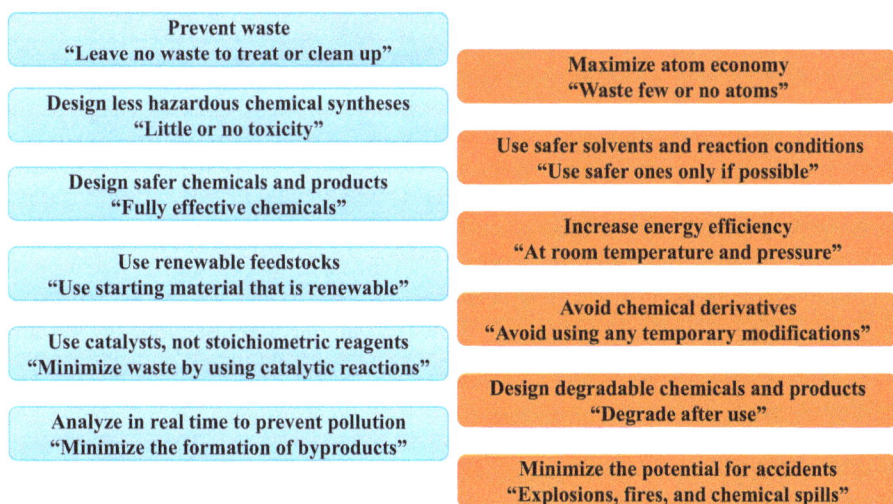

Prevent waste
"Leave no waste to treat or clean up"

Maximize atom economy
"Waste few or no atoms"

Design less hazardous chemical syntheses
"Little or no toxicity"

Use safer solvents and reaction conditions
"Use safer ones only if possible"

Design safer chemicals and products
"Fully effective chemicals"

Increase energy efficiency
"At room temperature and pressure"

Use renewable feedstocks
"Use starting material that is renewable"

Avoid chemical derivatives
"Avoid using any temporary modifications"

Use catalysts, not stoichiometric reagents
"Minimize waste by using catalytic reactions"

Design degradable chemicals and products
"Degrade after use"

Analyze in real time to prevent pollution
"Minimize the formation of byproducts"

Minimize the potential for accidents
"Explosions, fires, and chemical spills"

Figure 2.20: The twelve principles of green chemistry.

ducing the need for intensive heating or harsh chemical treatments. For example, the microbial fermentation process used to produce PHA occurs at moderate temperatures, consuming significantly less energy compared to traditional petrochemical-based polymer manufacturing.

2.5 Organic Polymers vs Inorganic Polymers

2.5.1 Definition

Organic polymers are large molecules made up of carbon-based monomers, typically containing carbon-hydrogen (C–H) bonds. They are usually derived from petroleum-based or natural sources. On the other hand, inorganic polymers, unlike organic polymers, do not rely on carbon-carbon (C–C) bonds. These polymers typically contain metal-oxygen (M–O), silicon-oxygen (Si–O), phosphorus-oxygen (P–O), or other non-carbon elements in their backbone structure.

2.5.2 Chemical Structure

Organic polymers are characterized by the presence of various functional groups such as hydroxyl, amine, carboxyl, and ester groups attached to the carbon backbone. They form strong covalent bonds between carbon atoms in the polymer chain, which provide stability and flexibility. In contrast to organic polymers, inorganic polymers are often based on silicon or phosphorus, with inorganic linkages like Si–O (silicone polymers) and P–N (polyphosphazene polymers).

2.5.3 Synthesis

Organic polymers are commonly synthesized through chain-growth polymerization methods, step-growth polymerization, and controlled polymerization. Alternatively, inorganic polymers like polyphosphazenes are typically synthesized via polycondensation reactions, often involving metal or silicon-based precursors, and silica-based inorganic polymers are made by the sol-gel process. Polymers with metal coordination centers are often synthesized through methods like metal-organic coordination or metal complexation.

2.5.4 Physical and Chemical Properties

Organic polymers, such as polyethylene and polypropylene, are known for being flexible, lightweight, and easy to mold into various shapes. They typically have lower melting points and are more prone to degradation at high temperatures. Some organic polymers are also sensitive to oxidation, UV radiation, and chemical degradation, although certain types are highly resistant to these factors. Generally, organic polymers act as insulators, but some conjugated polymers can display electrical conductivity. In contrast, inorganic polymers, like silicones and polysilanes, tend to have high thermal resistance, allowing them to perform at much higher temperatures than organic polymers. These polymers are generally more chemically stable, with some, like silicones, resisting acids, bases, and solvents. Certain inorganic polymers, especially those containing metal centers (e.g., polyphosphazenes), may also exhibit electrical or ionic conductivity. Additionally, many inorganic polymers have a more rigid structure, which can result in enhanced mechanical strength, particularly under high-temperature conditions.

2.5.5 Mechanical Properties

Organic polymers are typically lighter and more flexible than inorganic polymers, making them ideal for applications that require flexibility, stretchability, and impact resistance. However, they generally have lower tensile strength compared to many inorganic materials. In contrast, inorganic polymers tend to be more rigid and stronger, particularly at elevated temperatures. They are well-suited for structural applications that demand high-temperature stability and corrosion resistance. Additionally, inorganic polymers often possess high tensile strength, allowing them to withstand harsh environmental conditions.

2.5.6 Recycling

Organic polymers like PET and HDPE are recyclable, but others, such as PVC and PS, pose environmental challenges due to their complex processing requirements and potential toxicity. Moreover, many organic polymers are not biodegradable, leading to the accumulation of plastic waste.

On the other hand, inorganic polymers are typically more challenging to recycle, though certain types, such as silicones, can be repurposed through specialized methods. However, they generally have a lower environmental impact during degradation, as they do not usually produce toxic byproducts.

2.5.7 Organic Polymers

Organic polymers are large molecules made up of repeating units (monomers) primarily composed of carbon atoms, as well as hydrogen, oxygen, nitrogen, and other elements. They are widely used in various industries, including plastics, textiles, medicine, and electronics, due to their versatility and diverse properties. Here are some examples of organic polymers, their key characteristics, and common applications, summarized in Table 2.9.

Table 2.9: Examples of organic polymers.

Polymer	Monomer	Properties, types, and applications
Polyethylene (PE)	Ethylene	Its structure is either linear-branched or cross-linked. Low-density polyethylene (LDPE) is more flexible and is used in plastic bags and packaging films. High-density polyethylene (HDPE) is more rigid and dense, and it is used in containers, pipes, and toys. Ultra-high molecular weight polyethylene (UHMWPE) is a linear polyethylene with extremely long chains; it is incredibly tough, abrasion-resistant, impact-resistant, and used in high-performance areas like bulletproof vests, orthopedic implants, conveyor belts, and industrial machine parts. Ultra-low-molecular-weight polyethylene (ULMWPE) has very short polymer chains, is lightweight, flexible, has low strength, and is used in coatings, lightweight films, lubricants, and adhesives. High-density cross-linked polyethylene (HDXLPE) is a form of HDPE with crosslinked polymer chains, has high resistance to wear, impact, and chemical degradation, and is commonly used in medical implants, chemical storage tanks, and piping systems for high-stress environments. High-molecular-weight polyethylene (HMWPE) consists of long linear polyethylene chains, is tough, has excellent wear resistance, and good impact strength, and is used in industrial applications like conveyor belts, cutting boards, and automotive components.Cross-linked polyethylene (XLPE) has polyethylene chains chemically or physically crosslinked to form a three-dimensional network structure, is highly resistant to heat, chemicals, and stress cracking,

Table 2.9 (continued)

Polymer	Monomer	Properties, types, and applications
		and is widely used in electrical insulation, plumbing systems, and industrial pipes. Medium-density polyethylene (MDPE) has a linear structure with fewer branches than LDPE but more than HDPE, offers good toughness, flexibility, chemical resistance, and impact strength, and is used in gas pipes and water pipes. Linear low-density polyethylene (LLDPE) is a linear polymer with significant short-chain branching, is flexible, and has excellent tensile strength; it is used in stretch films, packaging materials, geomembranes, agricultural films, and flexible tubing. Chlorinated polyethylene (CPE) is a modified polyethylene where chlorine atoms replace hydrogen atoms in the polymer chain; it is flexible, weather-resistant, flame-retardant, and resistant to chemicals, and it is used in flexible hoses, roofing membranes, wire, and cable insulation.
Polystyrene (PS)	Styrene	Amorphous, with a benzene ring attached to a vinyl group. General-purpose polystyrene (GPPS) is transparent and rigid and is used in consumer goods. High-impact polystyrene (HIPS) is a blend of polystyrene with dispersed rubber; it is lightweight, impact-resistant, easy to process, and used in consumer products, packaging, toys, and electronics casings.
Polyisobutylene (PIB)	Isobutylene	It is a synthetic rubber polymer made from the polymerization of isobutylene, characterized by a highly branched, saturated hydrocarbon structure; it is highly elastic, impermeable to gases, resistant to chemicals and oxidation, and used in inner tubes, adhesives, sealants, and fuel additives.
Polypropylene (PP)	Propylene	It is a semi-crystalline thermoplastic polymer made from the polymerization of propylene, with a structure of repeating units arranged in isotactic, syndiotactic, or atactic configurations, and it is used in packaging (containers, films), automotive parts, textiles, and medical supplies.

Table 2.9 (continued)

Polymer	Monomer	Properties, types, and applications
Polyvinyl chloride (PVC)	Vinyl chloride	Linear or slightly branched polymer with chlorine atoms attached to the backbone, rigid or flexible (depending on additives), resistant to chemicals and weathering, and electrically insulating. Rigid PVC is used in pipes, window profiles, and flooring. Flexible PVC, made by adding plasticizers, is used in electrical cables and plumbing hoses.
Polyethylene terephthalate (PET)	Terephthalic acid and ethylene glycol	Semi-crystalline polymer with ester functional groups; strong, lightweight, resistant to moisture, high melting point, and recyclable. Amorphous PET is used in clear plastic bottles and films, while crystalline PET is used in applications such as fibers and containers.
Nylon (Polyamide, PA)	Diamines, such as hexamethylenediamine and dicarboxylic acids, such as adipic acid	Linear, with amide linkages between the repeating units; strong, tough, abrasion-resistant, and moisture-absorbent. Nylon 6,6 is made from hexamethylenediamine and adipic acid, while Nylon 6 is made from caprolactam. Both are commonly used in fabrics, fibers, textiles, and automotive parts.
Polyurethane (PU)	Diisocyanates, such as methylene diphenyl diisocyanate, and polyols	Contains urethane linkages in the polymer backbone, is elastic, strong, versatile, and can be rigid or flexible depending on the formulation. Rigid polyurethane is used in insulation foam, and flexible polyurethane is used in coatings and sealants.
Polytetrafluoroethylene (PTFE)	Tetrafluoroethylene	Linear polymer with fluorine atoms attached to the carbon backbone has extremely low friction, high chemical resistance, and high thermal stability, and is used in non-stick cookware coatings, gaskets, seals, electrical insulation, and bearings.
Polycarbonate (PC)	Bisphenol A (BPA) and phosgene	Contains carbonate linkages in the backbone, is transparent, tough, heat-resistant, and impact-resistant, can be molded easily, and is used in eyeglass lenses, optical discs, medical devices, safety equipment, and automotive parts.
Acrylonitrile-butadiene-styrene (ABS)	Acrylonitrile, butadiene, and styrene	A terpolymer with three distinct blocks is tough, impact-resistant, easily molded, has good chemical resistance, and is used in automotive parts, electronic housings, toys, and pipes.

Table 2.9 (continued)

Polymer	Monomer	Properties, types, and applications
Polyvinyl fluoride (PVF)	Vinyl fluoride	A thermoplastic polymer composed of repeating vinyl fluoride units, highly resistant to weathering, chemicals, and UV radiation, is used in protective coatings, solar panel films, industrial laminates, and aerospace components.
Polyetheretherketone (PEEK)	4,4'-Difluorobenzophenone and the disodium salt of hydroquinone	A high-performance thermoplastic polymer with an aromatic backbone containing alternating ether and ketone groups in its repeating units, it is extremely strong, lightweight, heat-resistant, and used in aerospace, automotive, medical implants, and electronics.
Polypropylene oxide (PPO) Also known as poly (propylene glycol)	Propylene oxide	Made by the polymerization of propylene oxide, with repeating units, it is flexible, lightweight, has good chemical and moisture resistance, and low thermal conductivity, and is used in foams, adhesives, and sealants.
Polyvinyl alcohol (PVA)	Polyvinyl acetate	A water-soluble synthetic polymer made by the hydrolysis of polyvinyl acetate and used in textiles, paper coatings, adhesives, water-soluble films, packaging, and as a stabilizer or thickener in cosmetics and pharmaceuticals.
Polyvinyl acetate (PVAc)	Vinyl acetate	A synthetic polymer composed of repeating vinyl acetate units with an amorphous structure, which is flexible, has good adhesion properties, is water-insoluble, and is widely used in adhesives, paints, coatings, paper coatings, and as a precursor to polyvinyl alcohol (PVA).
Polyacrylonitrile (PAN)	Acrylonitrile	A semi-crystalline polymer composed of repeating acrylonitrile units, which is strong, lightweight, chemically resistant, thermally stable, and has excellent barrier properties, is used in the production of carbon fibers, textiles, and filtration membranes.
Polyparaphenylene or poly(p-phenylene) (PPP)	p-Phenylene	A rigid, crystalline polymer composed of repeating para-linked benzene rings; it is thermally stable, chemically resistant, electrically conductive, and used in advanced materials such as conductive polymers, batteries, and sensors.

2.5.8 Inorganic Polymers

Inorganic polymers offer unique properties such as high thermal stability, flame resistance, electrical conductivity, and chemical resistance or resistance to environmental degradation, making them valuable in specialized applications across a variety of industries, including aerospace, electronics, construction, and medical devices. They differ from organic polymers in their backbone composition and often provide performance characteristics that organic polymers cannot match, especially in extreme environments or when high-performance materials are needed. Here are examples of inorganic polymers, along with their structures and applications (Table 2.10).

Table 2.10: Examples of inorganic polymers.

Polymer	Structure	Applications
Polysiloxanes	$\left(\!-O-\underset{\underset{R'}{\vert}}{\overset{\overset{R}{\vert}}{Si}}-\!\right)_n$	Used in construction, automotive, and electronics for their durability and resistance to weathering, as well as lubricants due to their low-friction properties.
Polysilanes	$\left(\!-\underset{\underset{R}{\vert}}{\overset{\overset{R}{\vert}}{Si}}-\!\right)_n$	Used in the electronics industry as a precursor material for semiconductors and thin films, as well as protective coatings for materials subjected to high temperatures or aggressive environments.
Polyphosphazenes	$\left(\!-N\!=\!\underset{\underset{Cl}{\vert}}{\overset{\overset{Cl}{\vert}}{P}}-\!\right)_n$	Used in fire-resistant coatings for textiles and construction, in the production of insulation materials for electronics, and in aerospace applications for their resistance to high temperatures and oxidation.
Polyborosiloxanes	$\underset{HO}{\overset{HO}{>}}B-O\!\left(\!\underset{\underset{CH_3}{\vert}}{\overset{\overset{CH_3}{\vert}}{Si}}-O\!\right)_n\!B\!\underset{OH}{\overset{OH}{<}}$	Used in high-temperature coatings for aerospace and military applications, these materials are also utilized as electrical insulators due to their excellent thermal and electrical insulating properties.

Table 2.10 (continued)

Polymer	Structure	Applications
Polyphosphoric acid		Used as a catalyst in the production of organic compounds and petroleum refining, as well as a corrosion inhibitor for protecting metals from rust and corrosion.
Polysilazanes		Used in the production of silicon nitride-based ceramics for high-temperature applications such as turbine engines, as well as fire-resistant and heat-resistant protective coatings.
Polycarbosilanes		Used as a precursor to silicon carbide ceramics, which are utilized in high-performance applications such as aerospace and electronics, as well as protective coatings for high-temperature environments, including turbine blades and engines.
Polysulfides		They are utilized as sealants and adhesives in the construction and aerospace industries for their durability and resistance to extreme conditions, as well as for waterproof coatings in construction and infrastructure to ensure effective waterproofing.

2.6 Essential Keywords

Addition polymerization A process in which monomers are added sequentially to a developing polymer chain without producing any byproducts. It usually involves monomers with unsaturated linkages, such as alkenes.

Amorphous polymers Are polymers without a well-organized molecular structure. These polymers are usually translucent and flexible but have less strength and heat resistance than crystalline polymers.

Biodegradable polymers Are those that can be degraded spontaneously by microorganisms. These polymers provide an environmentally acceptable alternative to typical plastics, which can remain in the environment for extended periods.

Biopolymers Are polymers made from renewable resources, typically derived from living organisms or produced through biosynthesis.

Blending The process of blending two or more polymers to produce a material with qualities that are a mixture of the original components.

Condensation polymerization A reaction that occurs when two or more different monomers react to form a polymer while eliminating a small molecule, such as water or alcohol.

Copolymers Are polymers made from two or more different types of monomers. The monomers can be arranged in different patterns, such as alternating, block, random, or grafted.

Copolymerization Is a crucial technique in polymer chemistry that allows for the creation of versatile polymers with custom-tailored properties.

Cross-linking Is the process of chemically bonding polymer chains together to form a network. This results in increased mechanical strength, thermal stability, and chemical resistance. Cross-linked polymers are usually more rigid and durable.

Crystallinity Refers to the degree to which polymer chains are aligned in a highly organized form. Polymers with increased crystallinity exhibit better mechanical characteristics and heat resistance.

Degree of polymerization (DP) The number of monomer units in a polymer chain. A greater DP typically leads to longer chains, which contribute to enhanced physical qualities, including strength and durability.

Elastomers Are polymers with elastic characteristics that can stretch and return to their original shape. Rubber is a commonly used example. These polymers are frequently crosslinked to enhance flexibility and robustness.

Glass transition temperature (Tg) The temperature at which a polymer changes from a hard, glassy state to a more flexible, soft state.

Green chemistry A field of chemistry focused on reducing negative environmental impacts by designing safer chemical processes and creating sustainable materials.

Inorganic polymers Are polymers that contain non-carbon elements, such as metal-oxygen (M–O), silicon-oxygen (Si–O), or phosphorus-oxygen (P–O), in their backbone structure.

Homopolymers Are polymers made from one type of monomer. They consist of identical repeating units.

Molecular weight Is the total weight of a polymer molecule, determined by the length of its polymer chains.

Monomers Are small building units that join together through polymerization to form polymers. Each monomer contains functional groups that enable it to bond with other monomers.

Organic polymers Are large molecules made up of repeating units (monomers), mostly consisting of carbon atoms, but also hydrogen, oxygen, nitrogen, and other elements.

Plasticizers Are additives that make polymers more flexible by lowering the intermolecular interactions between polymer chains.

Polymerization Is the process by which monomers chemically link to produce lengthy polymeric chains. The two main kinds of polymerization are addition and condensation polymerization.

Polymerization kinetics The study of polymerization processes, the factors influencing them, and the mechanisms by which polymers are formed.

Thermoplastics Polymers that soften when heated, allowing them to be molded or reshaped multiple times. They can be remelted and reprocessed without significant degradation.

Thermosets Polymers that undergo a permanent hardening process when heated and molded. Once set, they cannot be remelted or reshaped. These materials are typically strong, heat-resistant, and chemically stable.

2.7 Questions and Answers

Questions	Answers
1. What is polymerization?	Polymerization is the chemical process by which small molecules, called monomers, chemically bond together to form long-chain molecules known as polymers. This process can occur via two main mechanisms: addition polymerization and condensation polymerization.

(continued)

Questions	Answers
2. What is the difference between addition and condensation polymerization?	Addition polymerization is a process in which monomers with unsaturated bonds (such as alkenes) react to produce polymers. Condensation polymerization is a process in which two distinct monomers react, and a small molecule, such as water, is removed as a result.
3. What is the role of a catalyst in polymerization?	A catalyst is a substance that accelerates the polymerization process without being consumed in the reaction. In addition to polymerization, catalysts like free radicals, cations, or anions are used to initiate the reaction. In condensation polymerization, catalysts such as acids or bases are used to promote the reaction between the monomers.
4. What are the key differences between thermoplastic and thermosetting polymers?	Thermoplastics are polymers that can be melted and remolded when heated, making them recyclable. They usually have linear or branched structures. Thermosets are polymers that harden and rigidify as a result of a curing or cross-linking process during polymerization. Once set, they cannot be remelted or reshaped.
5. How do polymer chains affect the properties of a polymer?	Polymer properties are heavily influenced by the length and structure of their polymer chains. Longer chains typically result in polymers with higher strength, viscosity, and melting points. The structure (linear, branched, or cross-linked) also affects the material's properties, with cross-linked polymers generally exhibiting better heat resistance and mechanical strength.
6. What are the factors that determine the molecular weight of a polymer?	The molecular weight of a polymer is determined by the size of its polymer chains, which, in turn, is influenced by polymerization conditions such as time, temperature, and the concentration of monomers. The size and reactivity of monomers affect how long the polymer chains grow. The type and amount of catalyst used in the reaction can influence the polymerization rate and the length of the polymer chains.
7. What is cross-linking, and how does it affect the properties of a polymer?	The process of forming covalent bonds between polymer chains results in a network structure. This enhances the polymer's rigidity, thermal stability, and mechanical strength. Cross-linked polymers exhibit greater resistance to solvents, heat, and deformation.

(continued)

Questions	Answers
8. What are copolymers, and how do they differ from homopolymers?	Homopolymers are polymers made from a single type of monomer, such as polyethylene, which is made from ethylene monomers. Copolymers are polymers made from two or more different monomers. Copolymers often have improved properties compared to homopolymers, such as better flexibility or strength.
9. How do the properties of polymers change with temperature?	At low temperatures, polymers can become brittle and stiff, especially if they are glassy in nature. As the temperature increases, polymers become more flexible and may transition into a rubbery state. For thermoplastics, the polymer chains move more freely at higher temperatures, while for thermosets, the cross-linked structure remains stable at higher temperatures, providing strength and resistance.
10. What is the significance of the glass transition temperature (Tg) in polymers?	The temperature at which a polymer changes from a hard, glassy state to a soft, rubbery state. Below Tg, the polymer is rigid and brittle, whereas above Tg, it becomes flexible and more mobile. This property is essential for applications where polymers must withstand temperature fluctuations while maintaining specific levels of flexibility and toughness.
11. What are the four main types of copolymers?	Random, alternating, block, and graft.
12. Give three examples of polymerization process initiators.	Benzoyl peroxide, azobisisobutyronitrile, and boron trifluoride.
13. List three names of chain-transfer agents.	n-Octyl mercaptan, 1,8-dimercapto-3,6-dioxaoctane, and carbon tetrachloride.
14. Give three examples of the common cyclic monomers used in ring-opening polymerization (ROP).	Cyclic ethers, cyclic esters, and cyclic xanthates.
15. Why must monomers have bifunctionalities in condensation polymerization?	The reaction cannot proceed beyond dimer formation if the monomers lack bifunctional groups, as no additional bonding sites are available for further polymerization.

Chapter 3
Polymer Structure and Morphology

Polymers exhibit various morphologies, which refer to their structural forms and arrangements. Understanding polymer morphology is important, as it affects their mechanical, thermal, and optical properties, influencing their use in applications ranging from packaging materials to medical devices.

The arrangement of the polymer chains in any given material affects various properties such as crystallinity, transparency, and the ability to process the polymer. Therefore, understanding the structure and morphology of polymers is fundamental to designing and processing them for specific applications. The primary structure (monomer units and bonding) determines the polymer's basic chemical properties, while the secondary structure (crystalline/amorphous regions, chain orientation) governs mechanical, thermal, and optical behaviors. The tertiary and quaternary structures, including chain alignment and crosslinking, further fine-tune the polymer's properties. By manipulating these factors, manufacturers can tailor polymers for applications ranging from flexible to high-strength components.

3.1 Molecular Structure

The molecular structure of a polymer refers to the arrangement of atoms and bonds that form the polymer chain. This structure can be classified into different categories based on monomeric units, chain organization, and chain composition.

3.1.1 Monomeric Units

Polymers are composed of repeating units called monomers. The primary structure refers to the chemical structure of these repeating monomer units and the way they are connected in the polymer chain. For example, polyethylene is made from the monomer ethylene, while polystyrene is derived from styrene monomers (Figure 3.1).

3.1.2 Chain Structure

Polymer chains can adopt linear, branched, crosslinked, or network structures. Linear polymers consist of a single chain of monomer units with minimal branching or crosslinking. In contrast, branched polymers contain side chains or branches extending from the main backbone, leading to more complex physical properties, such as gelation or thicker films. Crosslinked polymers feature covalent bonds between chains,

https://doi.org/10.1515/9783111585734-003

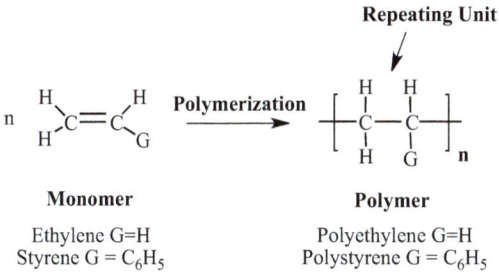

Figure 3.1: Examples of selected polymers and their monomeric units.

resulting in materials with increased rigidity and enhanced thermal stability. Network polymers, also known as polymer networks, are highly crosslinked structures where polymer chains are interconnected either directly or indirectly (Figure 3.2).

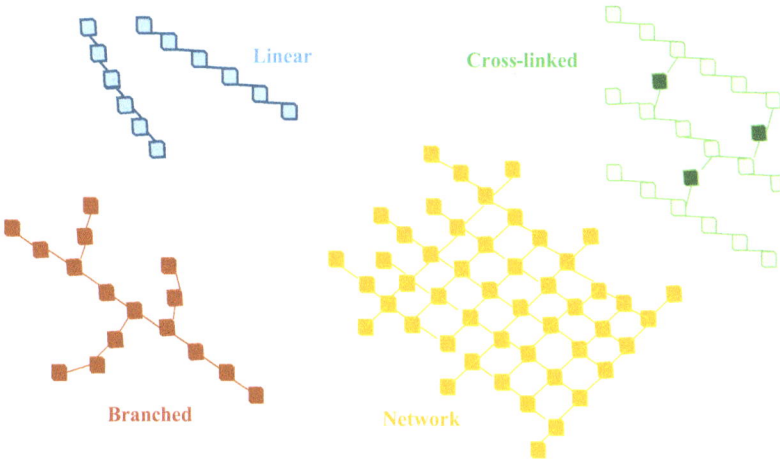

Figure 3.2: Polymer chain structures.

3.1.3 Polymer Crystallinity

Polymers can be amorphous, semi-crystalline, or crystalline depending on several factors, including repeating unit shapes, molecular weight, and thermal properties (Figure 3.3). Amorphous polymers lack a long-range ordered structure. Their molecules are arranged randomly, and there is no regular repeating pattern over large distances. They tend to be more flexible and have a broad glass transition temperature (Tg), above which they become soft and rubbery. On the other hand, crystalline polymers have ordered regions where the polymer chains pack in a regular, repeating pattern. However, they may still have amorphous regions. They tend to be more rigid, have higher melting temperatures, and their mechanical properties are usually superior in the crystalline regions. Semi-crystalline polymers exhibit both crystalline and amorphous regions.

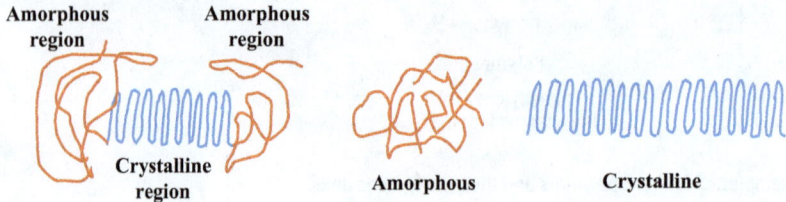

Figure 3.3: Amorphous, semi-crystalline, and crystalline polymer structures.

The morphology of semi-crystalline polymers can be complex, with crystallites embedded in an amorphous matrix. Spherical structures formed in the crystalline regions of semi-crystalline polymers during cooling are known as spherulites (Figure 3.4). These are spherical aggregates of lamellae that grow outward from a central nucleation point during crystallization. These spherulites influence the polymer's optical properties, making it opaque or translucent. Spherulites can be observed under a microscope, and their size and distribution can affect properties like transparency and mechanical performance. Smaller spherulites usually result in more transparent materials. Crystalline regions within the spherulites consist of thin layers called lamellae. These lamellae are made up of polymer chains that fold back on themselves in an organized manner. The formation of lamellae contributes to the mechanical properties of the polymer, such as tensile strength.

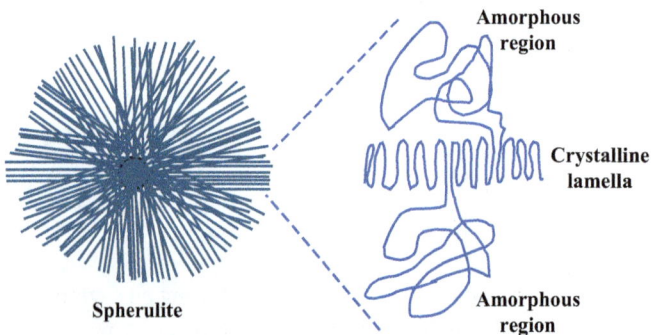

Figure 3.4: Different forms of polymer crystallinity.

3.2 Morphology of Polymer Blends and Composites

Polymer blends and polymer composites are indeed considered part of polymer morphology because their behavior and properties are heavily influenced by how their
components are organized at the microscopic or macroscopic level. The morphological features of these materials, such as phase separation, fiber orientation, particle
distribution, and the interface between different phases, directly impact the material's overall performance, making them an important aspect of polymer structure and
design.

3.2.1 Polymer Blends Morphology

Polymer blends refer to a mixture of two or more different polymers, which are combined to create a new material with properties derived from the individual components. The morphology of polymer blends refers to how the different polymer phases
(such as the dispersed phase and matrix phase) are arranged and interact with each
other at the microscopic level. Polymer blends can be either immiscible or miscible.
In immiscible blends, the two polymers do not form a homogeneous solution and instead form separate phases that are dispersed within one another. This phase separation can lead to microphase structures that affect the mechanical, thermal, and optical
properties of the blend. On the other hand, in miscible blends, the two polymers mix
at the molecular level to form a single phase. This results in a more homogeneous
structure, where the polymers are molecularly dispersed and mixed. The morphology
in miscible blends is often simpler than in immiscible blends, and the properties are
determined by the interactions between the polymer chains (Figure 3.5).

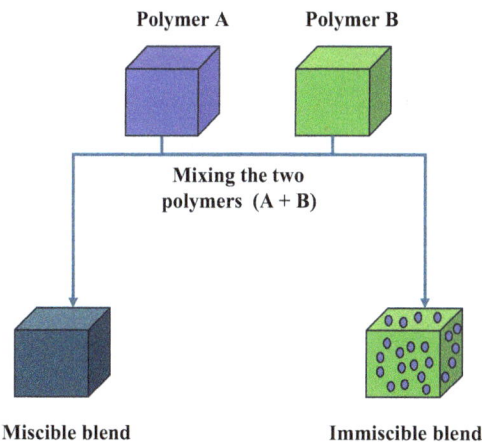

Figure 3.5: Miscible and immiscible blends.

3.2.2 Polymer Composites Morphology

Polymer composites are materials made by combining a polymer (the matrix) with a filler material (reinforcement), such as fibers (glass or carbon fibers), particles, or other materials, to improve certain properties, such as mechanical strength, thermal conductivity, and stiffness. The two main types of polymer composites are fiber-reinforced and particle-reinforced composites.

3.2.2.1 Fiber-Reinforced Composites

The most common type of polymer composite is fiber-reinforced polymer composites. The fiber orientation, fiber-matrix interaction, and fiber distribution determine the composite's morphology. Fiber alignment can be unidirectional, bidirectional, or random, and each alignment affects the material's strength, stiffness, and ductility. The composites can be continuously, discontinuously, or randomly aligned and can also be unidirectional, bidirectional, or multidirectional (Figure 3.6).

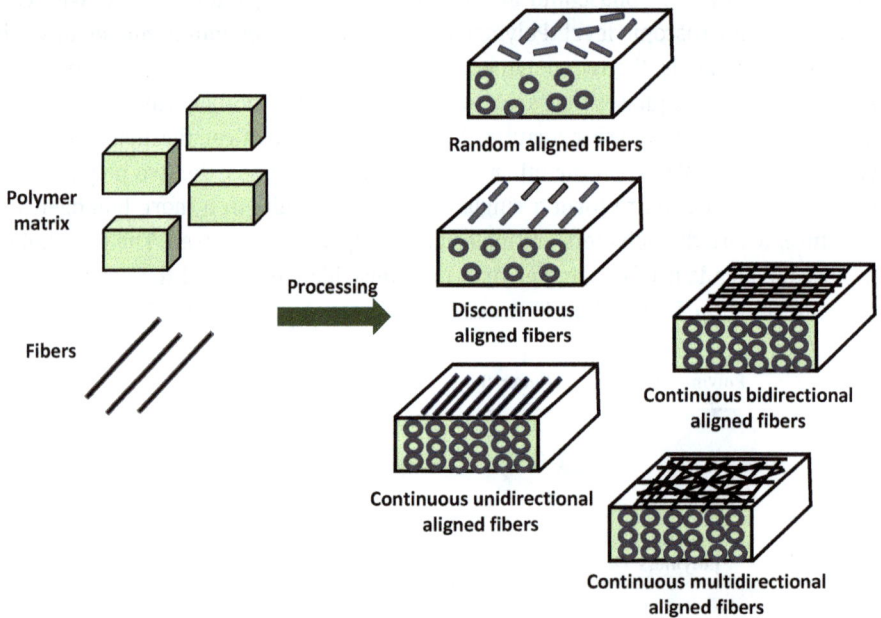

Figure 3.6: Fiber-reinforced composites.

3.2.2.2 Particle-Reinforced Composites

In these composites, particles or nanoparticles are incorporated into the polymer matrix to improve properties such as wear resistance, thermal stability, and impact strength. The morphology of particle-reinforced composites is determined by factors such as particle size, distribution, and the interaction between the particles and the matrix. Nanocomposites, in particular, utilize nanoparticles dispersed within the polymer matrix, resulting in distinctive morphological structures at the nanoscale (Figure 3.7).

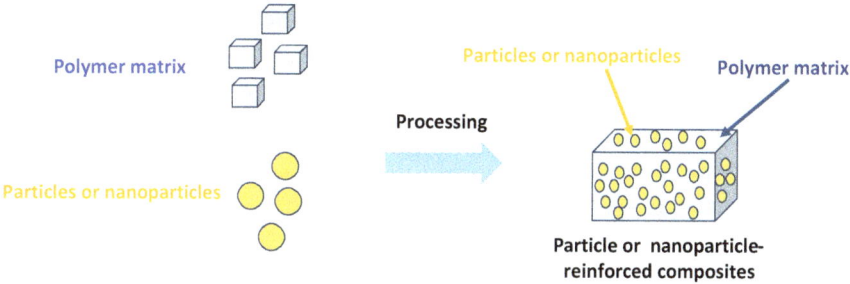

Figure 3.7: Nanoparticle or particle-reinforced composites.

3.3 Polymers Tacticity

Tacticity (Figure 3.8) plays a crucial role in polymer morphology, affecting crystallinity, molecular arrangement, and the resulting physical properties of the material. Isotactic and syndiotactic polymers typically form more ordered, crystalline structures, leading to enhanced strength, thermal resistance, and mechanical performance. In contrast, atactic polymers are generally more amorphous, providing increased flexibility but lower mechanical strength and crystallinity. Understanding a polymer's tacticity is essential for designing materials with specific properties for targeted applications. Depending on the arrangement of repeating monomer units along the polymer backbone, tacticity is classified into isotactic, syndiotactic, and atactic types.

In an isotactic polymer, all the side groups are oriented in the same direction along the polymer chain, resulting in a highly ordered structure. This regularity allows the polymer chains to pack closely together, leading to higher crystallinity. In a syndiotactic polymer, the side groups alternate between opposite sides of the polymer backbone. This alternating pattern can also lead to regular packing, but it is slightly less regular than isotactic arrangements. In an atactic polymer, the side groups are arranged randomly along the polymer chain. This irregular arrangement leads to a more disordered structure and typically results in amorphous (non-crystalline) regions in the polymer.

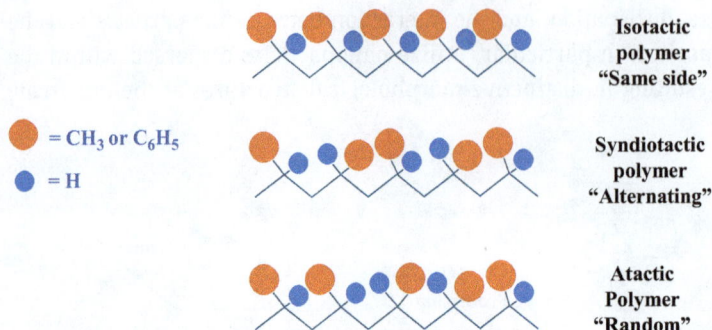

Figure 3.8: Polymer tacticity.

3.4 Polymers Head-to-Tail Configuration and Head-to-Head Configuration

Head-to-tail and head-to-head configurations in polymers are related to regio-regularity (regio-chemistry) rather than stereochemistry. However, they can influence the overall structural regularity of the polymer, which indirectly affects stereochemistry-related properties such as crystallinity and mechanical strength (Table 3.1). These terms describe how monomer units are connected during polymerization, particularly in substituted vinyl polymers such as polypropylene, polystyrene, and polyvinyl chloride (Figure 3.9).

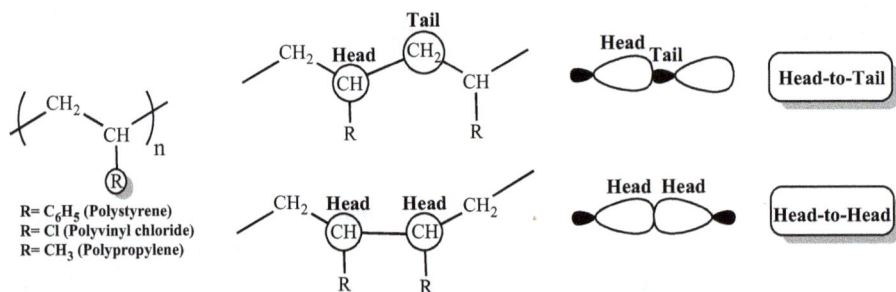

Figure 3.9: Polymers' head-to-tail configuration and head-to-head configuration.

Table 3.1: Comparison between polymers' head-to-tail configuration and head-to-head configuration.

Properties	Head-to-tail configuration	Head-to-head configuration
Stability and Regularity	Results in a more regular and orderly structure, which leads to increased polymer stability. This arrangement allows the polymer chains to pack closely together, leading to stronger intermolecular forces, such as van der Waals forces or hydrogen bonding, which can improve the polymer's physical properties.	Creates more irregular and less organized chains. These irregularities can cause weak points in the polymer structure, leading to reduced strength and stability.
Crystallinity	Tends to have higher crystallinity, meaning its chains are more aligned and able to form crystalline regions. Higher crystallinity typically results in better mechanical properties, such as increased tensile strength, hardness, and heat resistance.	Their arrangement disrupts the regular packing of polymer chains, which can reduce crystallinity. This often leads to a more amorphous polymer, which may be less durable and exhibit poorer mechanical properties.
Molecular Weight and Polymerization Efficiency	Usually more favorable for efficient polymerization, leading to higher molecular weight and longer polymer chains, which improve the material's overall performance.	Can create branching or irregular structures that interfere with the polymerization process, potentially resulting in shorter polymer chains and lower molecular weights.
Physical Properties	Exhibits better mechanical strength, toughness, and higher melting points due to its regular and orderly structure.	Being less ordered often results in lower mechanical strength, poorer chemical resistance, and reduced thermal stability.

3.5 Essential Keywords

Amorphous polymers Are molecules that lack a regular, repeating pattern in their molecular structure. These polymers tend to be more flexible, transparent, and less resistant to heat compared to crystalline polymers.

Branched polymers Are molecules that have side chains extending from the main polymer backbone.

Composites Are materials made by combining a polymer (the matrix) with a filler material (reinforcement), such as fibers (glass or carbon fibers), particles, or other materials, to improve certain properties, such as mechanical strength, thermal conductivity, and stiffness.

Cross-linked polymers Are molecules with covalent bonds that link different polymer chains together, creating a three-dimensional network. Cross-linking increases the rigidity, strength, and heat resistance of the polymer. Example: vulcanized rubber.

Crystalline polymers Are molecules that have ordered, repetitive structures in which the polymer chains align in a regular, repeating pattern. These polymers exhibit higher density, strength, and heat resistance.

Crystallinity Refers to the extent of ordered structures in polymer chains. High crystallinity results in stronger and more rigid polymers. Factors such as molecular weight, cooling rate, and monomer structure affect the degree of crystallinity.

Lamellar structure Crystallized chains in layers, with each layer consisting of ordered chains. The lamellae are typically surrounded by amorphous regions, which affect the overall properties of the polymer.

Linear polymers Are long chains of monomers connected end-to-end in a straight line.

Melting temperature (*Tm*) Temperature at which a crystalline polymer transitions from a solid state to a liquid. Crystalline polymers typically have a higher melting temperature than amorphous polymers.

Monomers The basic building blocks of polymers, consisting of small molecules that undergo polymerization to form long-chain structures. The structure and functional groups in the monomer affect the final properties of the polymer.

Morphology Describes the arrangement of polymer chains and their structure at various length scales, from the molecular level to the bulk material. It plays a key role in determining a polymer's mechanical, thermal, and optical properties.

Polymer blends Are combinations of two or more polymers designed to achieve specific properties that may not be possible with a single polymer alone.

A polymer chain Consists of repeating monomer units (mers) connected by covalent bonds. The arrangement and length of these chains have a significant impact on the polymer's overall properties.

Semi-crystalline polymers Are molecules that contain regions where the polymer chains are ordered (crystalline regions) and regions that are disordered (amorphous regions). These polymers balance flexibility and strength.

Spherulites Are spherical aggregates of polymer chains that form in semi-crystalline polymers during crystallization. These structures influence the optical and mechanical properties of the polymer. They are often observed under polarized light microscopy.

Tacticity Refers to the regularity of the arrangement of side groups (substituents) along the polymer chain.

3.6 Questions and Answers

Questions	Answers
1. What is the difference between crystalline and amorphous polymers?	Crystalline polymers have a highly ordered molecular structure, where polymer chains are aligned in a regular pattern. These polymers are typically stronger, denser, and have higher melting points. Amorphous polymers lack an ordered structure, resulting in a more random arrangement of polymer chains. These polymers are more flexible, transparent, and have lower melting points.
2. What is the benefit of knowing the glass transition temperature (Tg)?	Knowing the glass transition temperature (Tg) helps in determining a polymer's thermal limits, mechanical properties, and suitability for specific applications, ensuring optimal performance and processing conditions.
3. What are spherulites in semi-crystalline polymers?	Spherulites are spherical aggregates of crystalline polymer regions that form during the crystallization process of semi-crystalline polymers. These structures are often observed under polarized light microscopy and can influence the optical properties and mechanical performance of the polymer. The size and distribution of spherulites affect the material's transparency and strength.
4. What is the difference between isotactic, syndiotactic, and atactic polymers?	Isotactic polymers have all side groups on the same side of the polymer chain, leading to high crystallinity and rigidity. Syndiotactic polymers have side groups alternating sides, which can also allow for crystallinity but with different properties compared to isotactic polymers. Atactic polymers have side groups arranged randomly along the chain, leading to lower crystallinity and higher amorphousness, which makes them more flexible.
5. What is polymer crystallinity, and how does it affect properties?	Crystallinity refers to the degree of ordered structure in a polymer's molecular arrangement. High crystallinity typically results in increased strength, rigidity, and thermal resistance, as the polymer chains are closely packed in an organized structure. Amorphous regions, however, provide flexibility. Crystallinity influences properties such as clarity, mechanical strength, and heat resistance.

(continued)

Questions	Answers
6. How does the molecular weight of a polymer affect its morphology?	A higher molecular weight leads to longer polymer chains, which can result in better mechanical properties like strength and toughness, as the polymer chains have more entanglements. This can increase the crystallinity if the chains are able to align well. However, for amorphous polymers, higher molecular weight generally leads to improved elasticity and resistance to deformation.
7. What are the key factors affecting polymer morphology?	Cooling rate: Rapid cooling tends to favor the formation of amorphous structures, whereas slow cooling allows for greater crystallinity. Molecular weight: Longer chains tend to increase crystallinity, but excessively long chains can hinder crystallization. Polymer additives: Plasticizers, stabilizers, and other additives can influence the ordering of polymer chains. Processing conditions: Temperature and pressure during processing can significantly impact polymer morphology.
8. What is a lamellar structure in polymers?	The lamellar structure refers to a layered arrangement of crystalline polymer regions that form during crystallization. These layers are typically separated by amorphous regions and contribute to the material's overall strength, transparency, and thermal resistance. This structure is common in semi-crystalline polymers.
9. How do polymer blends affect polymer morphology?	When different polymers are blended together, the resulting morphology can depend on their compatibility: Compatible blends tend to mix well, forming a homogeneous structure. Incompatible blends may form phase-separated structures, where the individual polymers retain distinct regions within the material. The morphology of polymer blends directly impacts properties such as toughness, flexibility, and chemical resistance.

(continued)

Questions	Answers
10. What is the role of cross-linking in polymer morphology?	Cross-linking chemically bonds polymer chains into a three-dimensional network, improving rigidity, thermal stability, and chemical resistance. Unlike thermoplastics, cross-linked polymers, such as vulcanized rubber, cannot be re-melted. The extent of cross-linking affects elasticity and durability.
11. Give an example of a common polymer composite and its application.	Carbon fiber-reinforced polymer (CFRP) is commonly used in aerospace and automotive industries because of its excellent strength-to-weight ratio, which makes it ideal for lightweight yet strong applications.
12. Why are polymer composites stronger than pure polymers?	Because the reinforcement materials, such as fibers or nanoparticles, enhance the material's mechanical strength, stiffness, and durability, they play a crucial role in improving its overall performance.
13. What are the two main types of polymer blends?	The two main types are miscible and immiscible blends.
14. How do polymer blends differ from polymer composites?	Polymer blends are mixtures of two or more polymers, whereas polymer composites consist of a polymer matrix reinforced with fillers such as fibers or nanoparticles to enhance properties.
15. What are the benefits of polymer blends?	Polymer blends are used to enhance material properties such as toughness, flexibility, chemical resistance, and processability for applications in packaging, automotive, electronics, and medical devices.

Chapter 4
Polymer Characterization and Testing

Polymer characterization and testing are essential for evaluating polymers' structure, properties, and performance in real-world applications. Through a combination of chemical analysis, molecular weight determination, thermal testing, and mechanical testing, it is possible to gain detailed insights into how polymers behave under different conditions. These tests are crucial for ensuring the polymers' quality, reliability, and suitability for their intended use, whether in consumer goods, engineering materials, or specialized applications like aerospace and medical devices.

4.1 Polymer Characterization Techniques

Polymer characterization encompasses a wide range of techniques to analyze the structure, molecular weight, thermal properties, rheological behavior, and chemical composition of polymers. These methods provide insights into how polymers behave and perform under different conditions.

4.1.1 Chemical Structure Analysis

4.1.1.1 Fourier Transform Infrared Spectroscopy (FTIR)

Fourier Transform Infrared Spectroscopy (FTIR) is a reliable and cost-effective analytical technique used for polymer identification and assessing the quality of plastic materials. When a plastic sample absorbs infrared light, typically in the mid-infrared region, it produces a unique spectral pattern (either absorbance or transmittance), serving as a distinctive "fingerprint" for easy screening and analysis across various applications. FTIR provides insights into the chemical functional groups and overall molecular structure of a polymer. By measuring the absorption of infrared radiation, it detects vibrational changes in specific bonds within the polymer. The resulting spectrum reveals the presence of various functional groups, including amines, amides, alcohols, phenols, alkanes, carboxylic acids, aldehydes, ketones, alkenes, primary amines, aromatics, esters, ethers, alkyl halides, and aliphatic amines. This technique is widely used to identify polymer types, detect contaminants or impurities, and analyze polymer blends or copolymers. Figure 4.1 illustrates a simplified diagram of the main FTIR components.

https://doi.org/10.1515/9783111585734-004

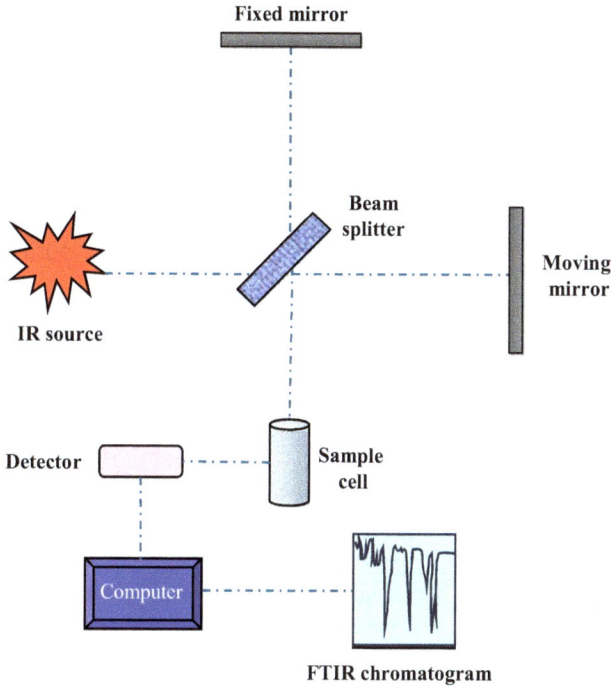

Figure 4.1: Schematic diagram of the FTIR's main components.

4.1.1.2 X-Ray Fluorescence (XRF)

X-ray fluorescence (XRF) is a highly effective analytical technique for both qualitative and quantitative elemental analysis of materials. It is particularly useful for measuring film thickness and composition, determining elemental concentrations in solids and solutions by weight, and detecting specific and trace elements within complex sample matrices. XRF analysis is widely applied across various industries, including plastics, rubber, textiles, fuels, chemicals, and environmental analysis. This method is fast, precise, non-destructive, and typically requires minimal sample preparation. Figure 4.2 presents a simplified diagram of the main XRF components.

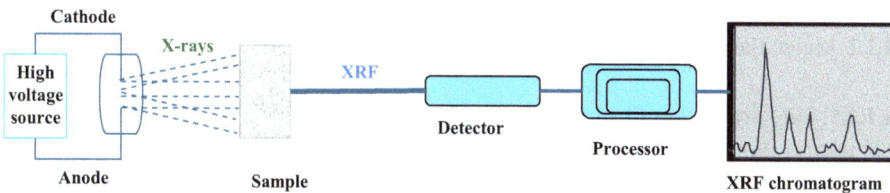

Figure 4.2: X-ray fluorescence main components.

4.1.1.3 Nuclear Magnetic Resonance (NMR) Spectroscopy

Nuclear Magnetic Resonance (NMR) spectroscopy is a powerful technique for analyzing the molecular structure of polymers, particularly the arrangement of atoms and the types of monomer units. By examining the magnetic properties of atomic nuclei, NMR provides detailed insights into the chemical structure, molecular connectivity, and conformational characteristics of a polymer. This technique is essential for determining molecular weight distribution, polymer tacticity (such as isotactic or syndiotactic configurations), and copolymer composition. Figure 4.3 illustrates a simplified schematic of the main components of an NMR system.

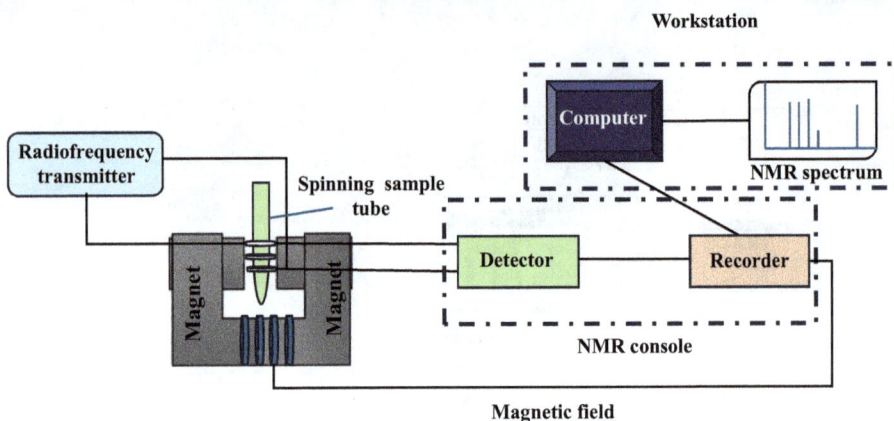

Figure 4.3: The schematic diagram of the NMR.

4.1.1.4 X-Ray Diffraction (XRD)

X-ray diffraction (XRD) is an analytical technique used to examine the crystalline structure of polymers and assess their degree of crystallinity. In this method, X-rays interact with the crystal planes of a polymer, producing a diffraction pattern that reveals details about its crystallinity and molecular packing arrangement. XRD is a valuable tool for understanding polymer morphology and structural properties. Figure 4.4 presents a simplified diagram of the main components of an XRD system.

4.1.2 Molecular Weight Analysis

4.1.2.1 Gel Permeation Chromatography (GPC)

Gel permeation chromatography (GPC), also known as size-exclusion chromatography (SEC) or gel-filtration chromatography (GFC), is a technique that separates molecules based on their effective size in the mobile phase.

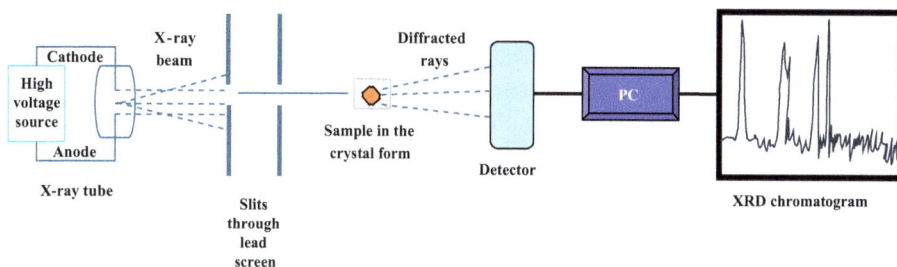

Figure 4.4: X-ray diffraction's main components.

For organic material analysis, the chromatography column is packed with a highly cross-linked spherical polystyrene/divinyl benzene matrix with a precisely controlled pore diameter. Molecules smaller than the pore size can diffuse in and out of the pores, while larger molecules are excluded. As a result, larger molecules elute more rapidly than smaller ones (Figure 4.5). Unlike other chromatographic techniques, GPC relies on a mechanical separation process rather than a chemical interaction.

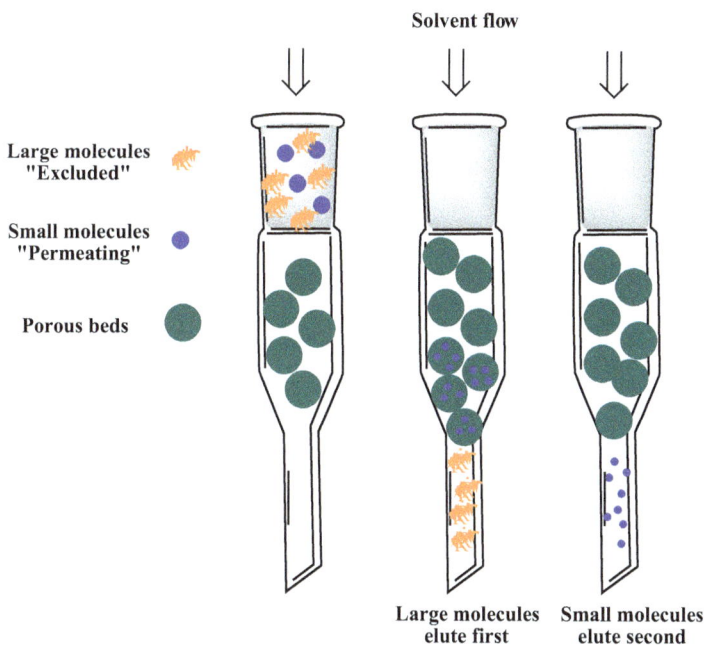

Figure 4.5: Gel permeation chromatography principle.

4.1.2.2 Viscosity Measurements

The viscosity measurement of polymer solutions in organic solvents provides a value that is closely related to the polymer's molecular mass. Generally, higher viscosity indicates a higher molecular weight, although the correlation is not strictly linear. The test involves dissolving the polymer in a suitable solvent at a very low concentration. Both the pure solvent and the polymer solution pass through a capillary, and their flow times are recorded. Two common types of viscometers used for fluid measurements are the Ostwald and Ubbelohde viscometers.

The Ostwald viscometer (Figure 4.6A) is a laboratory instrument made of a glass capillary tube with a bulb at one end, where the liquid sample is drawn into the capillary through suction. The viscosity is determined by measuring the time required for a fixed volume of liquid to flow through the capillary under gravity. This viscometer is particularly well-suited for transparent liquids and is widely employed in the chemical, pharmaceutical, and food industries.

The Ubbelohde viscometer (Figure 4.6B) is a common instrument for measuring the viscosity of liquids and operates on the principle of capillary flow. It consists of a vertically oriented, thin-walled capillary tube with a bulb at the bottom. The liquid sample is drawn into the capillary by suction, and the time it takes for a specific volume to flow through the tube is measured. This viscometer is particularly suitable for transparent liquids with low to medium viscosity, such as oils and polymers.

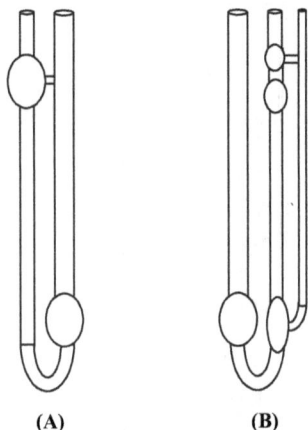

(A) (B)
Ostwald viscometer **Ubbelohde viscometer** **Figure 4.6:** Ostwald and Ubbelohde viscometers.

Traditional glass capillary viscometers require manual flow time measurement using a stopwatch, demanding precise attention and accuracy from the technician or operator. Additionally, reliable thermostat baths and appropriate bath liquids are essential for accurate results. To improve efficiency and reduce resource consumption, automatic gravimetric capillary viscometers have been developed. These advanced models can automatically fill, measure, clean, and dry the sample, eliminating human error.

They also feature special capillaries that work with smaller measuring volumes than the original types. These automatic viscometers typically include liquid baths, sometimes coupled with thermoelectric systems, for precise temperature control. In addition to gravity-driven capillary viscometers, pressurized devices are also used. Pressure can be applied in three ways: with weight, a motor, or gas pressure. Common examples include weight-controlled capillary viscometers, rotational viscometers, and gas-pressurized capillary viscometers (Figure 4.7).

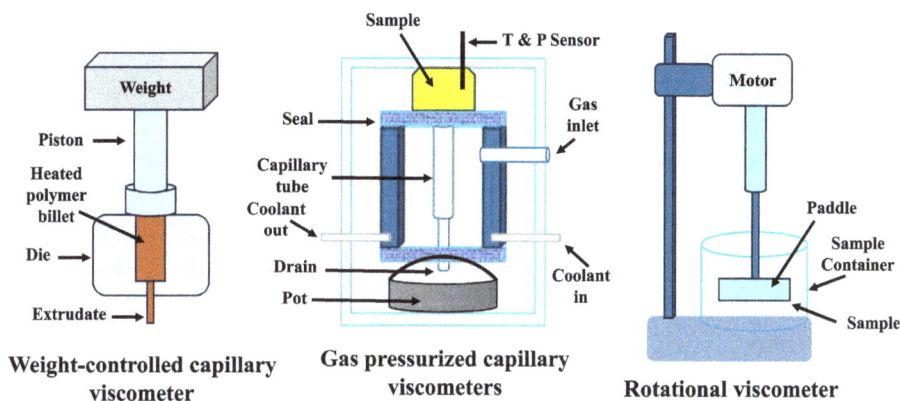

Figure 4.7: Pressurized capillary viscometers.

In a weight-controlled capillary viscometer, a defined weight placed on top of a piston is pulled downward by gravity. As the steel piston moves inside a vertical steel cylinder containing the sample, it forces the sample through an extrusion die (capillary) at the bottom of the cylinder. Rotational viscometers are widely used for measuring the viscosity of various substances with high accuracy. In this method, viscosity is determined by measuring the torque (shearing stress) applied to the cylindrical surface of a rotor immersed in the sample while rotating at a constant speed. Gas-pressurized capillary viscometers utilize a glass or steel capillary with a precisely defined inner diameter (ranging from 0.2 mm to 1 mm) and length (typically between 30 mm and 90 mm). In these viscometers, a controlled gas pressure forces the sample through the capillary at a preset rate, ensuring precise and consistent viscosity measurements.

4.1.3 Thermal Analysis

4.1.3.1 Differential Scanning Calorimetry (DSC)

Differential Scanning Calorimetry (DSC) tracks heat changes associated with phase transitions and chemical reactions in relation to temperature variations. It measures the difference in heat flow between a sample and a reference as the temperature fluctuates.

When any given sample is heated from room temperature to its decomposition temperature, peaks with positive and negative $\Delta dH/dt$ values appear, each corresponding to a specific thermal event, such as melting or crystallization.

In summary, DSC (Figure 4.8) detects heat flow into or out of a polymer sample during heating or cooling, providing vital information about its thermal transitions, such as melting point, glass transition temperature, and crystallization behavior.

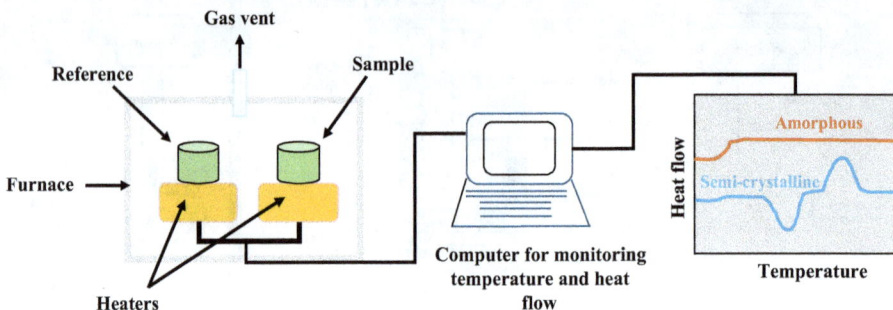

Figure 4.8: Schematic representation of the DSC.

4.1.3.2 Thermogravimetric Analysis (TGA)

Thermogravimetric Analysis (TGA) is a thermal analysis technique that measures changes in the physical and chemical characteristics of materials as they vary with temperature or time. It measures the mass change of a polymer as it is heated at a constant rate or maintained at a constant temperature. By continuously monitoring the sample's mass, TGA provides insights into thermal stability and decomposition behavior. Mass loss typically indicates material degradation due to vaporization, oxidation, or decomposition. This technique is particularly valuable for determining key parameters such as degradation temperature, decomposition kinetics, and the presence of additives in polymers. A schematic representation of TGA is shown in Figure 4.9.

4.1.3.3 Dynamic Mechanical Analysis (DMA)

Dynamic Mechanical Analysis (DMA) is a powerful technique for studying and characterizing materials, particularly their viscoelastic behavior. It measures key viscoelastic properties of polymers, including storage modulus, loss modulus, and damping behavior, providing valuable insights into their mechanical performance under varying conditions. In Dynamic Mechanical Analysis (DMA), a sample is typically subjected to oscillatory stress while its strain response is measured. This allows for the determina-

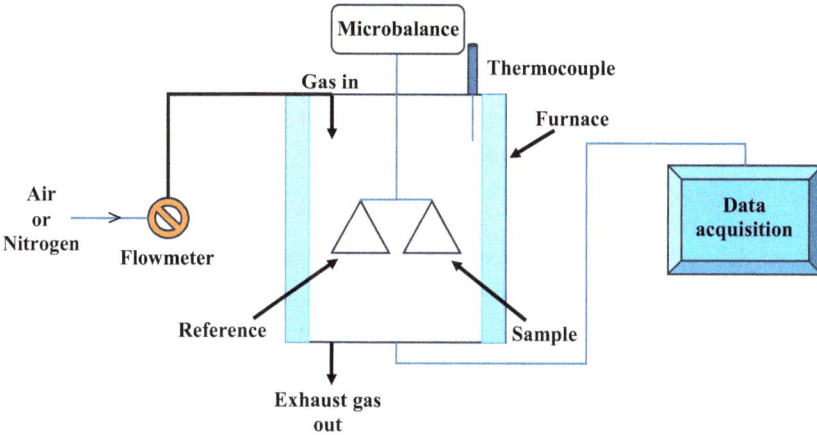

Figure 4.9: Schematic representation of the TGA.

tion of key viscoelastic properties, including the storage modulus (elastic component) and loss modulus (viscous component), which provide insights into the material's mechanical behavior. This analysis enables the measurement of mechanical characteristics such as phase transitions (e.g., glass transition) as well as the influence of temperature or frequency on polymer behavior. Figure 4.10 shows the essential components of the DMA mechanical instrument.

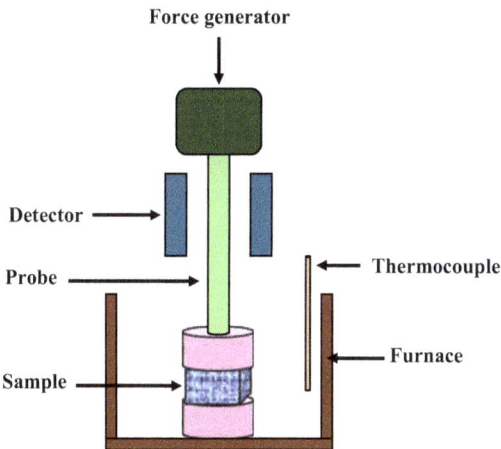

Figure 4.10: Basic components of the DMA.

4.1.4 Morphology Analysis

4.1.4.1 Scanning Electron Microscopy (SEM)

Scanning Electron Microscopy (SEM) is an effective method for capturing high-resolution images of a polymer's surface morphology. It operates by scanning an electron beam across the sample's surface and capturing the reflected electrons to create detailed images. This technology enables the imaging of a wide range of materials, including polymers, and provides vital information about their surface structure and texture.

SEM is also fast, unrestrictive, and requires minimal sample preparation. The approach examines polymer surface characteristics, microstructure, and fracture surfaces while also providing a high-resolution, three-dimensional visualization platform for in-depth analysis and characterization of micro and nanoscale polymer surfaces and internal structures. Figure 4.11 shows a schematic illustration of the SEM.

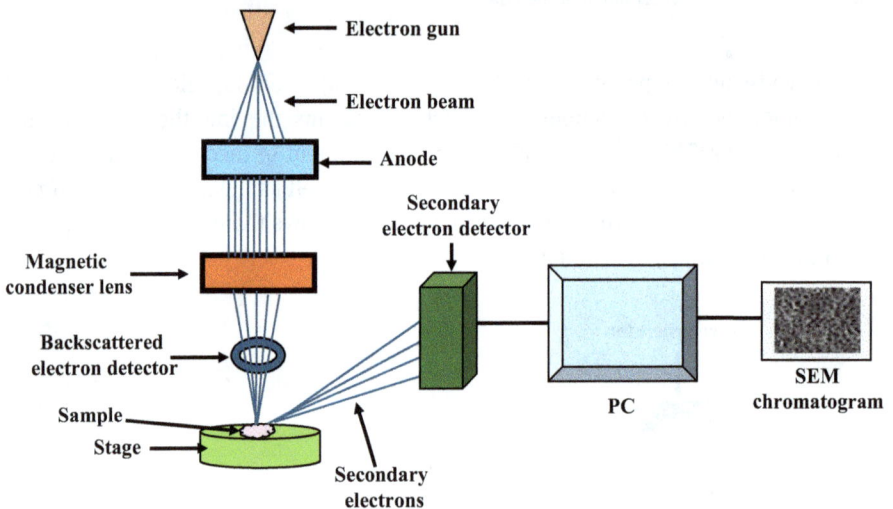

Figure 4.11: Schematic representation of the SEM.

4.1.4.2 Atomic Force Microscopy (AFM)

Atomic Force Microscopy (AFM) provides detailed information on surface topography at the nanoscale. A sharp tip scans the surface of the polymer, and the interaction between the tip and the surface is measured to create a 3D topographic map. It is used to analyze the surface roughness, morphology, and mechanical properties of polymer materials and is capable of quantifying the surface roughness of samples down to the angstrom scale. This flexible technique can be used to obtain high-resolution nanoscale images and study local sites in air (conventional AFM) or liquid (electrochemical AFM) environments. The basic components of the AFM system are presented in Figure 4.12.

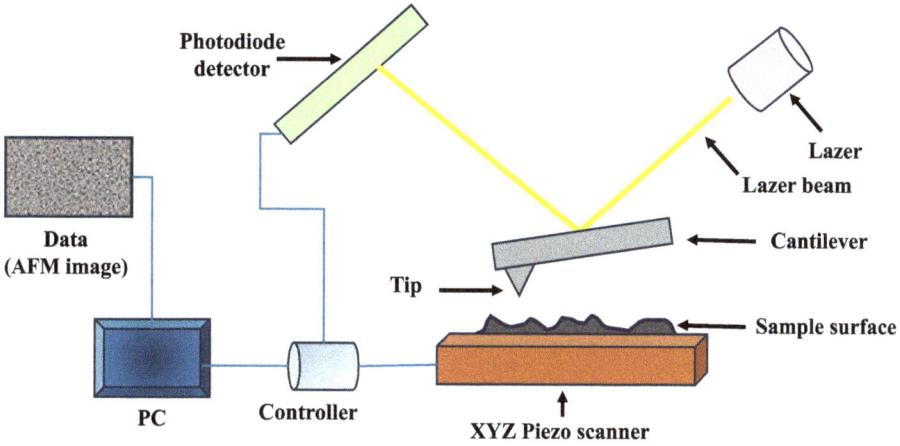

Figure 4.12: Schematic representation of the AFM.

4.1.4.3 Transmission Electron Microscopy (TEM)

Transmission Electron Microscopy (TEM) is a high-resolution imaging technique in which an electron beam passes through a thin sample to reveal its internal structure at the nanoscale level. In a TEM system (Figure 4.13), electrons travel through the material, and the resulting image is captured on a fluorescent screen or a digital detector.

TEM offers better resolution than visible light because electrons have a shorter wavelength, allowing for the study of atomic and molecular-level features. This makes

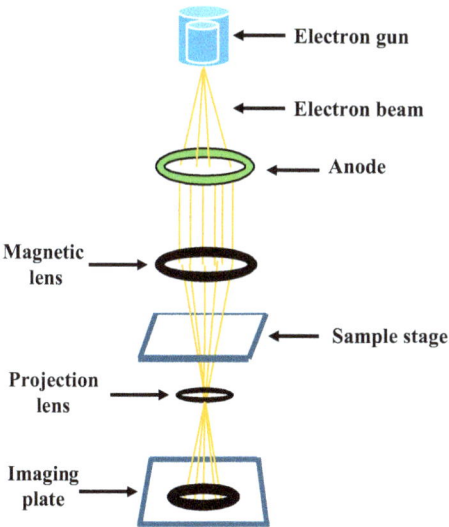

Figure 4.13: Schematic representation of the TEM.

TEM an effective tool for investigating microstructures, material composition, and crystallography in a wide range of substances.

4.2 Polymer Testing Techniques

Polymer testing involves measuring the mechanical, thermal, optical, rheological, and chemical properties of polymers to evaluate their performance in real-world applications. These tests simulate the stresses, strains, and environmental conditions that the polymer will face in service.

4.2.1 Mechanical Testing

4.2.1.1 Tensile Testing

Tensile testing measures the polymer's response to elongation and determines properties such as tensile strength, elongation at break (fracture strain), and Young's modulus. A polymer specimen is stretched until it breaks, and the stress-strain curve is recorded. The stress is plotted against the strain to determine key mechanical properties. This test is generally used to assess the strength, flexibility, and elasticity of polymers. A simple schematic diagram is shown in Figure 4.14.

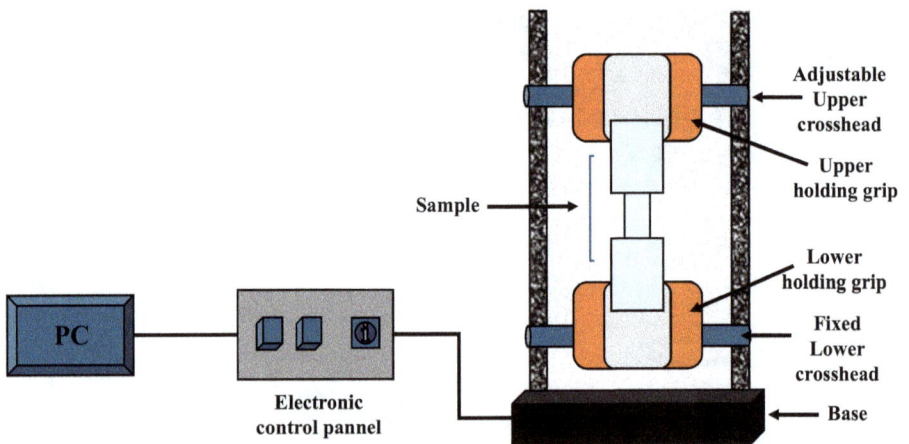

Figure 4.14: Schematic diagram of the tensile testing machine.

4.2.1.2 Impact Testing (Izod/Charpy)

Impact testing (Figure 4.15) evaluates a polymer's toughness and resistance to sudden forces (e.g., fractures or cracks). A hammer strikes the polymer specimen, and the amount of energy absorbed before breaking is measured. By this testing method, the resistance of polymers to brittle fracture and impact damage is determined.

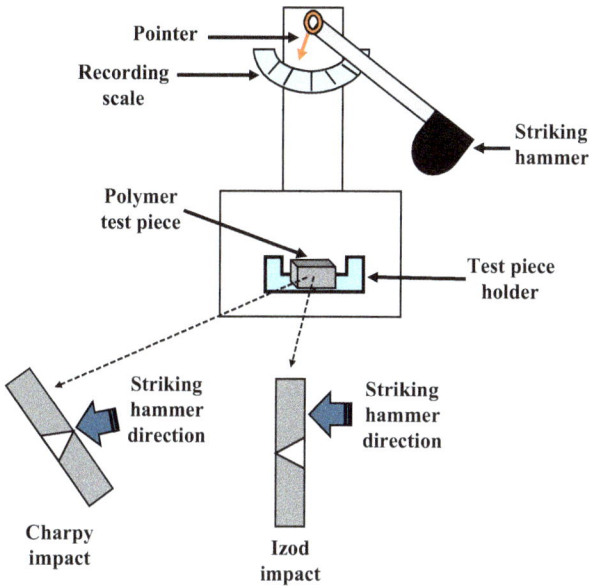

Figure 4.15: Schematic diagram of the impact-testing machine.

4.2.1.3 Hardness Testing

Hardness testing (Figure 4.16) measures the resistance to indentation of a polymer surface. A hard indenter is pressed into the polymer surface under a specified load, and the size or depth of the indentation is measured. This method is typically used to assess the hardness of polymers in applications where surface wear or deformation is significant (e.g., coatings or tires).

4.2.1.4 Flexural Testing

Flexural testing (Figure 4.17) measures the bending strength and stiffness of polymers. A polymer sample is bent under a known force, and the maximum stress and strain at the point of failure are recorded.

4.2.2 Rheological Testing

Rheological testing evaluates the flow behavior and viscoelastic properties of polymers, including melt viscosity, shear thinning, and the relationship between stress

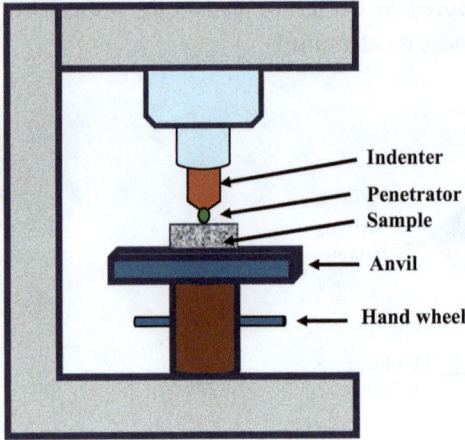

Figure 4.16: Schematic diagram of the hardness-testing machine.

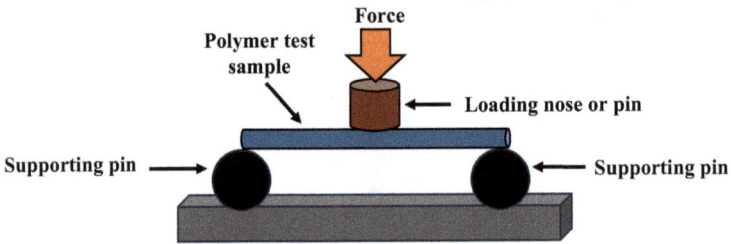

Figure 4.17: Schematic diagram of the flexural testing machine.

and strain. A polymer sample is subjected to shear forces using instruments like a rheometer (Figure 4.18), where the relationship between the applied stress and the resulting strain (or flow) is measured.

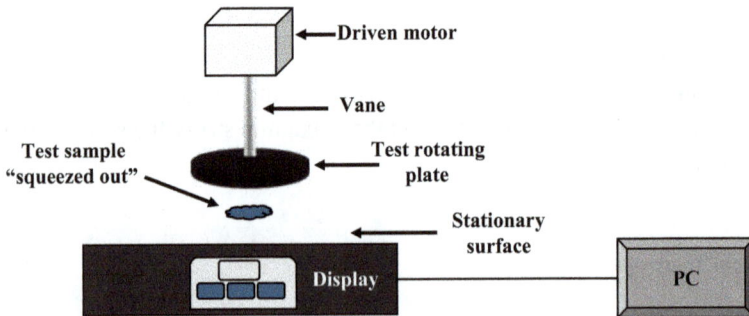

Figure 4.18: Schematic diagram of the rheometer.

4.2.3 Environmental and Chemical Testing

4.2.3.1 Weathering and UV Stability

These tests measure how a polymer material performs when exposed to weathering conditions such as UV light, heat, and humidity. This testing method is commonly used in the automotive, construction, and outdoor products industries to ensure that polymers maintain their properties under long-term exposure to outdoor conditions. Figure 4.19 shows a schematic diagram for testing weathering resistance and UV stability.

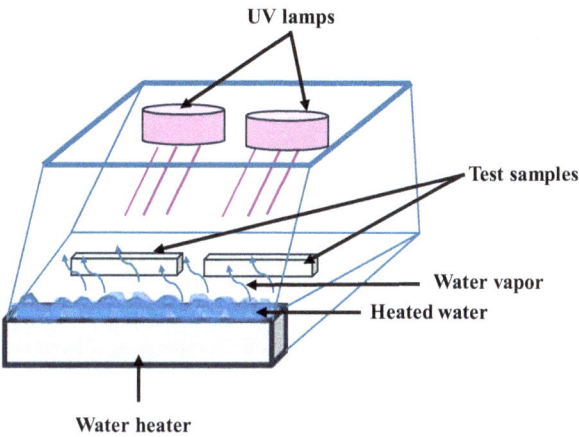

Figure 4.19: Schematic diagram of the UV and weathering resistance test chamber.

4.2.3.2 Chemical Resistance Testing

This test evaluates a polymer's resistance to degradation when exposed to various chemicals, such as solvents, acids, or bases (Figure 4.20). It is important for materials used in chemical processing, medical devices, and packaging to resist chemical corrosion. Chemically resistant polymers help maintain mechanical properties when exposed to various chemicals. They prevent degradation, cracking, or softening, leading to longer product life. They not only ensure product safety but also prevent material failure. There are two common testing methods, namely immersion and soak testing.

Immersion testing involves submerging a material sample in a liquid for a set duration under controlled conditions, such as temperature, pressure, and humidity. The purpose is to assess the material's ability to resist chemical degradation, absorb or resist water, and withstand exposure to corrosive substances without losing its properties or mechanical strength. This testing helps determine the material's durability and performance when exposed to various environmental factors.

Soak testing is used to evaluate the chemical resistance of materials, including plastics, elastomers, and composites. This test involves immersing a material sample in a chemical solution for a specified period, typically under controlled conditions of temperature, pressure, and humidity. The sample is then evaluated for changes in its physical and mechanical properties, such as weight gain or loss, color changes, and cracking or crazing.

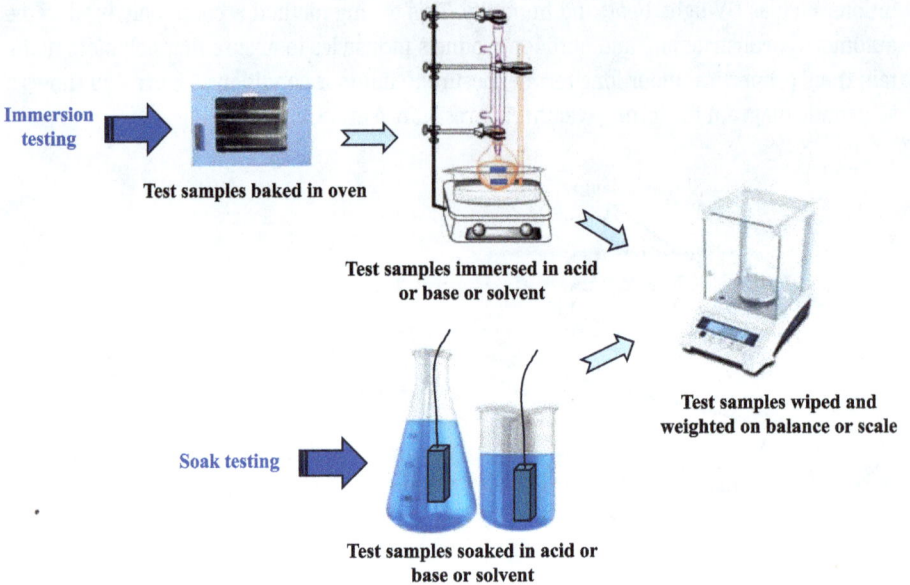

Figure 4.20: Schematic diagram of the chemical resistance testing.

Polymers used in the plastics industry vary in their ability to withstand chemical exposure, with some offering superior resistance, while others are more vulnerable. Table 4.1 presents a selection of polymers and their respective durability and resistance characteristics.

Table 4.1: Examples of polymers and their durability or resistance profiles.

Polymer type	Superior durability or protection	Limited durability or protection
High-density polyethylene	– Acids: Dilute and concentrated acids, such as sulfuric acid, hydrochloric acid, and nitric acid. – Bases: Strong bases, such as sodium hydroxide and potassium hydroxide. – Alcohols: Methanol, ethanol.	– Strong Oxidizers: Concentrated sulfuric acid (>70%), fuming nitric acid, and peroxides. – Aromatic and Chlorinated Solvents: Toluene, benzene, xylene, chloroform, carbon tetrachloride. – Halogenated Compounds: Bromine, chlorine gas.

Table 4.1 (continued)

Polymer type	Superior durability or protection	Limited durability or protection
Low-density polyethylene	– Acids: Dilute and moderately concentrated acids, such as hydrochloric acid and sulfuric acid. – Bases: Sodium hydroxide and potassium hydroxide. – Alcohols: Methanol, ethanol, and isopropanol.	– UV Radiation: Degrades over time without stabilizers. – High Temperatures: Softens above approximately 85 °C, limiting high-temperature applications. – Strong Oxidizers: Fuming nitric acid and concentrated sulfuric acid. – Aromatic and Chlorinated Solvents: Benzene, toluene, xylene, chloroform.
Polylactic acid	– Organic Acids: Mild concentrations of lactic acid, acetic acid, and citric acid. – Aqueous Solutions: Water, saline, weak alkaline solutions.	– Strong Acids and Bases: Hydrochloric acid, sulfuric acid, and sodium hydroxide. – Solvents: Acetone, benzene, toluene, and chloroform. – High Temperatures: Softens at around 60 °C, making it unsuitable for heat-resistant applications.
Polypropylene	– Acids: Sulfuric acid (up to 80%), hydrochloric acid (HCl, any concentration), phosphoric acid (H_3PO_4, any concentration), and acetic acid (any concentration). – Bases: Strong alkalis (sodium hydroxide, potassium hydroxide). – Alcohols: Methanol, ethanol, isopropanol.	– Strong Oxidizers: Concentrated sulfuric acid ($\geq$80%), fuming nitric acid. – Aromatic and Chlorinated Solvents: Benzene, toluene, xylene, chloroform. – Halogenated Compounds: Bromine, chlorine gas. – UV Radiation: Prone to degradation unless UV stabilizers are added.
Polystyrene	– Water and Moisture: Does not absorb water, making it ideal for packaging. – Weak Acids and Bases: Dilute acetic acid, citric acid, and mild alkaline solutions. – Alcohols: Ethanol, isopropanol (moderate resistance). – Oils and Greases: Good resistance to many oils and greases.	– Strong Acids and Bases: Concentrated sulfuric acid, nitric acid, and sodium hydroxide. – Aromatic and Chlorinated Solvents: Benzene, toluene, xylene, chloroform.

Table 4.1 (continued)

Polymer type	Superior durability or protection	Limited durability or protection
Polyvinyl chloride	– Acids: Dilute and moderate concentrations of hydrochloric acid, sulfuric acid, and nitric acid. – Bases: Moderate resistance to sodium hydroxide and potassium hydroxide. – Alcohols: Good resistance to ethanol, methanol, and isopropanol. – Water and Aqueous Solutions: Highly resistant; does not absorb water.	– Strong oxidizing agents, such as concentrated nitric acid and chlorine bleach, can degrade PVC. – Aromatic hydrocarbons, such as benzene and toluene, can cause swelling and stress cracking in PVC. – Halogenated hydrocarbons, such as chloroform and carbon tetrachloride, can also cause swelling and stress cracking.
Acrylonitrile butadiene styrene	– Water and Moisture: Does not absorb water, making it suitable for outdoor and humid environments. – Weak Acids and Bases: Dilute hydrochloric acid, acetic acid, citric acid, and mild alkaline solutions. – Oils and Greases: Exhibits good resistance to mineral oils, greases, and lubricants. – Alcohols: Ethanol, isopropanol (moderate resistance). – Impact and Wear: High toughness and scratch resistance compared to other plastics.	– Strong Acids and Bases: – Concentrated sulfuric acid, nitric acid, and sodium hydroxide. – Aromatic and Chlorinated Solvents: Benzene, toluene, xylene, chloroform. – High Temperatures: Softens around 105 °C, limiting its use in high-heat environments.
Polytetrafluoroethylene (Teflon)	– Acids (Strong and Weak): Hydrochloric acid, sulfuric acid, nitric acid, hydrofluoric acid. – Bases (Strong and Weak): Sodium hydroxide, potassium hydroxide, ammonia. – Solvents: Acetone, benzene, toluene, chloroform, ethyl acetate, alcohols. – Oils and Greases: Excellent resistance to petroleum-based and synthetic oils. – Water and Moisture: Completely hydrophobic; it does not absorb water. – Oxidizers: Hydrogen peroxide, sodium hypochlorite (bleach), ozone. – High Temperatures: Maintains stability up to 260 °C and can withstand short-term exposure to approximately 316 °C.	– Molten Alkali Metals: Sodium, potassium (reacts at high temperatures) – Fluorine gas and some fluorinated compounds: Can degrade under extreme conditions.

Table 4.1 (continued)

Polymer type	Superior durability or protection	Limited durability or protection
Polyetheretherketone	– High Temperatures: Maintains dimensional stability and strength up to 250–300 °C. – Acids (Strong and Weak): Hydrochloric acid, sulfuric acid, nitric acid, phosphoric acid (moderate to high resistance). – Bases (Strong and Weak): Sodium hydroxide, potassium hydroxide, ammonia. – Solvents: Resistant to a wide range of solvents, including acetone, ethanol, toluene, and oils.	– Halogenated Compounds: Chlorine, bromine (degrade under long-term exposure). – Strong Oxidizers (Concentrated): Fuming sulfuric acid, fuming nitric acid, and concentrated nitric acid (may cause degradation over time). – Alkali Metals and High-Pressure Fluorine: Can react under certain conditions. – UV Radiation: Can degrade without UV stabilizers or protective coatings, though it is less sensitive than some other plastics.
Nitrile rubber (NBR)	– Oils and Fuels: Petroleum-based oils, diesel, gasoline, hydraulic fluids. – Solvents: Aliphatic hydrocarbons, such as hexane and heptane. – Water and Aqueous Solutions: Exhibits good resistance to water and salt solutions. – Acids and Bases: Dilute hydrochloric acid, sulfuric acid, and sodium hydroxide.	– UV and Ozone Exposure: Prone to cracking without stabilizers. – Extreme Temperatures: Becomes brittle below − 30 °C. – Degrades at temperatures above 100–120 °C without special formulations. – Strong Acids and Bases: Concentrated sulfuric acid, nitric acid. – Strong Alkalis: Potassium hydroxide, sodium hydroxide.
Silicone rubber	– Water and Moisture: Completely waterproof; does not absorb water. – Oils and Greases: Resistant to many mineral and synthetic oils. – Alcohols and Aqueous Solutions: Methanol, ethanol, isopropanol, and saline solutions – Acids and Bases: Dilute acids such as acetic acid and citric acid, and weak alkalis. – UV Radiation and Ozone: Excellent resistance; does not degrade like other rubbers.	– Strong Acids and Bases: Concentrated sulfuric acid, nitric acid, and hydrochloric acid. – Strong alkalis, such as sodium hydroxide and potassium hydroxide. – Aromatic and Chlorinated Solvents: Benzene, toluene, xylene, chloroform. – Ketones and Esters: Acetone, MEK, Ethyl Acetate.

Table 4.1 (continued)

Polymer type	Superior durability or protection	Limited durability or protection
Polyurethane	– Solvents (Some): Moderate resistance to certain solvents such as acetone, alcohols (ethanol, isopropanol), and esters. – Acids and Bases (Mild to Moderate): Dilute acetic acid, sulfuric acid, hydrochloric acid, and sodium hydroxide.	– Strong Acids and Bases: Concentrated sulfuric acid, nitric acid, and hydrochloric acid. – Strong alkalis: Potassium hydroxide, sodium hydroxide. – Aromatic and Chlorinated Solvents: Benzene, toluene, xylene, chloroform. – Ketones: Acetone, MEK. – UV Radiation and Ozone: Tend to degrade materials with prolonged exposure, leading to brittleness and discoloration unless UV stabilizers are added. – High Temperatures: Flexible PU starts to lose its properties at temperatures above 80–100 °C. – Rigid PU can maintain stability up to 120–150 °C.
Melamine formaldehyde	– Water and Moisture: Highly resistant. – Acids: Weak to moderate, acetic acid, citric acid. – Bases: Dilute ammonia, mild alkaline cleaners. – Solvents: Alcohols, esters, and hydrocarbons. – Heat and Fire: High heat resistance and self-extinguishing properties.	– Strong Acids and Bases: Concentrated sulfuric acid, hydrochloric acid, and nitric acid. – Strong Alkalis: Sodium hydroxide, potassium hydroxide. – Aromatic and Chlorinated Solvents: Benzene, toluene, chloroform.
Urea formaldehyde	– Water and Aqueous Solutions: Resistant to moisture but not fully waterproof. – Weak Acids: Dilute acetic acid, citric acid. – Oils and Greases: Generally resistant to most non-aggressive oils and greases.	– Strong Acids and Bases: Hydrochloric acid, sulfuric acid, and nitric acid. – Sodium hydroxide, potassium hydroxide. – UV Radiation: Becomes brittle with prolonged exposure.

4.3 Essential Keywords

Crystallinity Refers to the degree of order in a polymer's molecular structure. Polymers with high crystallinity are generally stronger, more rigid, and have higher melting points.

Differential Scanning Calorimetry (DSC) Is a technique used to study the heat flow associated with polymer transitions, such as melting, crystallization, and the glass transition. It provides insights into the thermal behavior and stability of polymers.

Fourier Transform Infrared Spectroscopy (FTIR) Is a technique used to identify functional groups and chemical bonds in a polymer. FTIR offers detailed information about the polymer's chemical structure and is valuable for analyzing additives, fillers, and degradation products.

Gel permeation chromatography (GPC) A commonly used technique to measure molecular weight distribution.

Glass transition temperature (Tg) The temperature at which an amorphous polymer transitions from a rigid, glassy state to a flexible, rubbery state. It is a critical property that affects the thermal performance of polymers.

Hardness testing A technique that measures a polymer's resistance to surface indentation.

Impact testing (Izod or Charpy tests) evaluates a polymer's ability to withstand sudden or high-impact forces without breaking. It is important for assessing a material's toughness and brittleness under real-world conditions.

Melting temperature (Tm) The temperature at which crystalline polymers transition from a solid to a liquid. It indicates the thermal stability and the processing conditions required for crystalline polymers.

Molecular weight Refers to the average size of the polymer chains. It influences the polymer's strength, viscosity, and processability.

Nuclear Magnetic Resonance (NMR) Spectroscopy Is a technique used to study the chemical structure, monomer sequence, and functional groups of polymers. Both Proton (^{1}H) and Carbon (^{13}C) NMR are commonly used to identify the arrangement of monomers along the polymer chain.

Rheology Is the study of the flow and deformation behavior of materials. Rheological testing, using rheometers or viscometers, measures viscosity, elasticity, and flow behavior, providing essential insights into the processing properties of polymers during manufacturing.

Scanning Electron Microscopy (SEM) Is a method that provides high-resolution images of a polymer's surface morphology. It is particularly useful for examining polymer fracture surfaces, surface roughness, and the distribution of fillers or additives within the material.

Tensile testing A method that determines the force needed to stretch a polymer sample until it breaks. It offers critical information on the polymer's tensile strength, elongation, and modulus, which are required to understand its mechanical properties.

Thermogravimetric Analysis (TGA) Is a method that analyzes the change in mass of a polymer as it is heated, offering insights into its thermal stability, degradation temperature, and composition.

Transmission Electron Microscopy (TEM) Is a sophisticated imaging method used to investigate morphological characteristics at the nanoscale. It is used to study the internal structure of polymers, such as phase separation and crystalline domain layout.

X-ray diffraction (XRD) A technique used to study the crystallinity and morphology of polymers. It helps identify the ordered structures in crystalline polymers and provides information about the polymer's interatomic spacing and degree of crystallinity.

4.4 Questions and Answers

Questions	Answers
1. What is the purpose of polymer characterization?	Polymer characterization is the process of assessing a polymer's physical, chemical, and mechanical characteristics to better understand its structure, performance, and suitability for specific applications. Spectroscopy, chromatography, and microscopy are useful for determining molecular structure, molecular weight, crystallinity, and thermal behavior.
2. What is the role of molecular weight in polymer characterization?	Molecular weight is a critical parameter in polymer characterization, as it directly impacts the polymer's mechanical, thermal, and processing properties. Polymers with higher molecular weight typically demonstrate improved strength, toughness, and heat resistance. Techniques such as gel permeation chromatography (GPC) are commonly employed to determine the molecular weight distribution of polymers, providing insights into the polymer's size and structure.

(continued)

Questions	Answers
3. How is the molecular structure of a polymer analyzed?	A polymer's molecular structure can be studied using techniques such as Nuclear Magnetic Resonance (NMR), which provides information about the chemical structure and sequence of monomers in the polymer chain; Fourier Transform Infrared Spectroscopy (FTIR), which identifies functional groups in the polymer; and X-ray diffraction (XRD), which examines crystallinity and polymer morphology.
4. What is Differential Scanning Calorimetry (DSC), and how is it used in polymer testing?	Differential Scanning Calorimetry (DSC) is a method for measuring the heat flow associated with temperature-dependent polymer transitions. It is used to calculate the glass transition temperature (Tg), melting temperature (Tm), crystallization temperature, and heat capacity of polymers, which are critical for understanding their thermal behavior.
5. What is the significance of Thermogravimetric Analysis (TGA) in polymer testing?	Thermogravimetric Analysis (TGA) measures the change in mass of a polymer sample as a function of temperature. This technique is essential for determining the thermal stability, degradation temperature, and composition of polymers, and it helps evaluate how polymers behave under heat, providing valuable information about their ability to withstand high temperatures without decomposing. By analyzing the mass loss during heating, TGA also sheds light on the polymer's degradation mechanisms and the presence of additives or other components within the material.
6. How is the mechanical strength of a polymer tested?	The mechanical strength of a polymer is typically evaluated using tests such as tensile testing, which measures elongation and the force required to break a polymer sample; impact testing, which evaluates the polymer's resistance to sudden stress or fracture; and hardness testing, which assesses the polymer's resistance to indentation.
7. What is the importance of rheology in polymer characterization?	Rheology is the study of the flow and deformation behavior of materials. In polymer characterization, rheological testing measures the viscosity, elasticity, and flow behavior of polymers, which are crucial for processing and shaping polymers.

(continued)

Questions	Answers
8. How is polymer crystallinity determined?	Polymer crystallinity is determined using techniques such as X-ray diffraction (XRD), which provides information on the ordered, crystalline regions within the polymer; differential scanning calorimetry (DSC), which identifies melting and crystallization behaviors; and polarized light microscopy, which allows observation of crystalline structures such as spherulites in semi-crystalline polymers.
9. What is the function of Scanning Electron Microscopy (SEM) in polymer testing?	Scanning Electron Microscopy (SEM) provides high-resolution images of a polymer's surface morphology. It is used to examine the surface features, fracture patterns, and microstructure of polymers, which help in analyzing failure modes, surface roughness, and the distribution of additives or fillers within the polymer matrix.
10. How are polymer degradation and weathering tested?	Polymer degradation and weathering are tested by subjecting the polymer to accelerated aging conditions, such as UV exposure, temperature cycling, and exposure to moisture or chemicals.
11. Which thermoplastic can withstand all chemicals, including extreme acids and oxidizers?	Polytetrafluoroethylene (PTFE).
12. What are the two common chemical resistance testing methods?	Immersion and soak testing.
13. What does rheological testing do?	Rheological testing evaluates the flow behavior and viscoelastic properties of polymers, including melt viscosity, shear thinning, and the relationship between stress and strain.
14. Name three polymer morphology analysis techniques.	Scanning Electron Microscopy (SEM), Atomic Force Microscopy (AFM), and Transmission Electron Microscopy (TEM).
15. What are the two types of viscometers that can be used for fluid measurements?	Ostwald and Ubbelohde viscometers.

Chapter 5
Polymer Processing Technologies

Polymer processing involves a range of techniques, where the selection of a specific method depends on factors such as polymer type (thermoplastic or thermoset), intended application, and desired final properties. Each technique has distinct advantages, enhancing production efficiency, material performance, and cost-effectiveness. Given the extensive use of polymers in industries such as automotive, medical, and consumer goods, polymer processing plays a vital role in advancing materials technology and meeting industry demands. This process converts raw polymer materials into finished products through various manufacturing methods that shape, mold, or form polymers into functional items. As a key aspect of polymer engineering, it significantly impacts the final properties and overall performance of polymer-based products. Below is an in-depth overview of common polymer processing techniques:

5.1 Molding

5.1.1 Injection Molding

Injection molding (Figure 5.1) is a common method for producing complex, high-precision components from thermoplastic polymers. It is ideal for large-scale manufacturing of products, including automobile components, medical equipment, packaging, and consumer goods. Polymer pellets are heated and melted within the barrel of an injection molding machine. The molten polymer is then injected under high pressure into a closed mold cavity, where it cools and solidifies into the required form. Once completely solidified, the finished item is ejected from the mold and is ready for further processing or use.

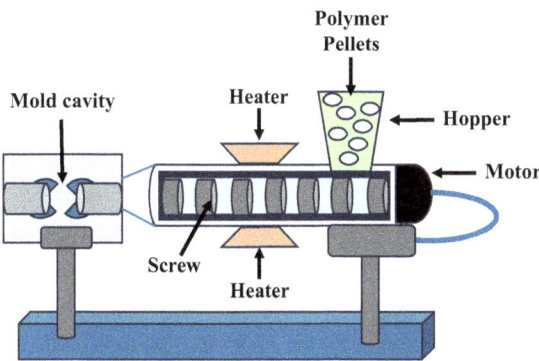

Figure 5.1: Schematic diagram of the Injection molding machine.

https://doi.org/10.1515/9783111585734-005

5.1.2 Blow Molding

Blow molding (Figure 5.2) is used to create hollow polymer products, such as bottles, containers, and tanks, typically from thermoplastic materials. The polymer is melted and either extruded into a tube (parison) or injected into a mold, and then inflated with air, causing it to conform to the shape of the mold. The formed product is cooled and ejected from the mold.

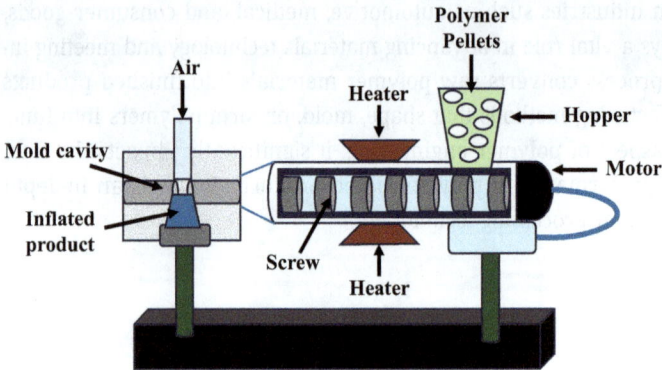

Figure 5.2: Schematic diagram of the blow-molding machine.

5.1.3 Extrusion Molding

Extrusion molding is a continuous manufacturing process that converts polymers into diverse shapes such as sheets, films, and pipes. In this process (Figure 5.3), polymer pellets or powder are fed into an extruder barrel, where they are heated and melted using thermal energy and mechanical shear. The molten polymer is then pressed through a die, resulting in the desired form. After separating from the die, the produced polymer is cooled using air, water, or chilled rollers. The finished product is either cut to specified lengths or coiled into rolls for further processing or distribution.

5.1.4 Rotational Molding

Rotational molding (Figure 5.4) is used to produce hollow plastic products, such as large tanks, containers, and playground equipment, typically made from thermoplastics. During the process, polymer powder is placed into a hollow mold. The mold is then heated while rotating on two perpendicular axes, allowing the polymer to melt and coat the interior surface of the mold. After the polymer has melted and fully coated the mold, it is cooled and solidified. The finished product is then removed from the mold.

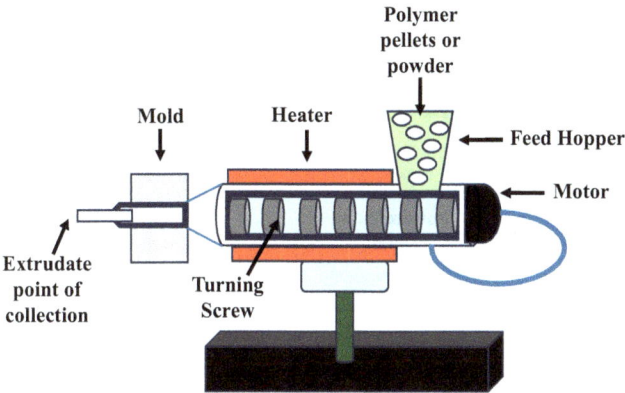

Figure 5.3: Schematic diagram of the extrusion molding machine.

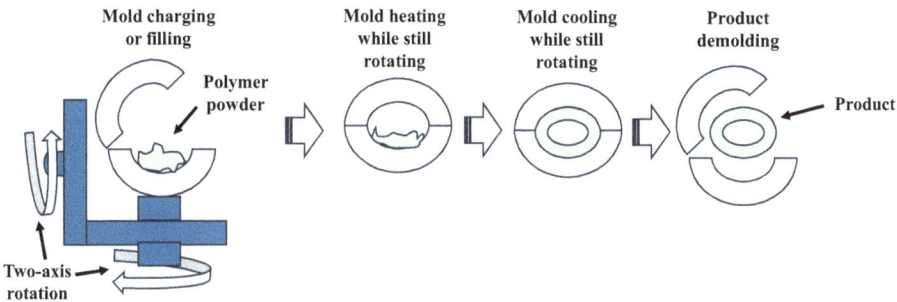

Figure 5.4: Schematic diagram of the rotational molding machine.

5.1.5 Compression Molding

Compression molding (Figure 5.5) is a widely used technique for shaping thermosetting polymers and rubber materials. It is particularly suitable for manufacturing automotive components, electrical insulators, and large structural parts. In this process, a predetermined amount of polymer is placed into a mold, which is then closed. Heat and pressure are applied, causing the material to flow and take the shape of the mold. Once the polymer has cured, the molded part is removed, completing the process.

5.2 Thermoforming

Thermoforming (Figure 5.6) involves heating a thermoplastic sheet until it becomes soft and moldable, then shaping it into the desired form using a mold. During this process, a polymer sheet is heated until it softens and becomes pliable. The sheet is either vacuum-

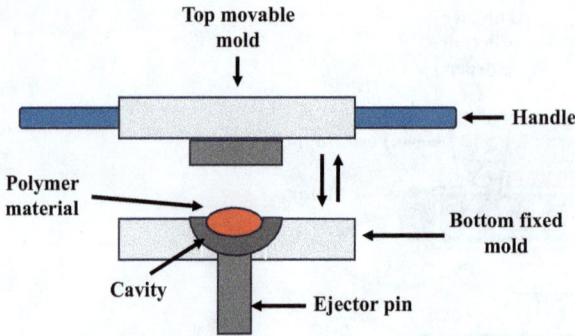

Figure 5.5: Schematic diagram of the compression molding machine.

formed (using vacuum pressure) or pressure-formed (using air or mechanical pressure) into a mold. The formed part is cooled to solidify the shape, and the excess material is then trimmed away.

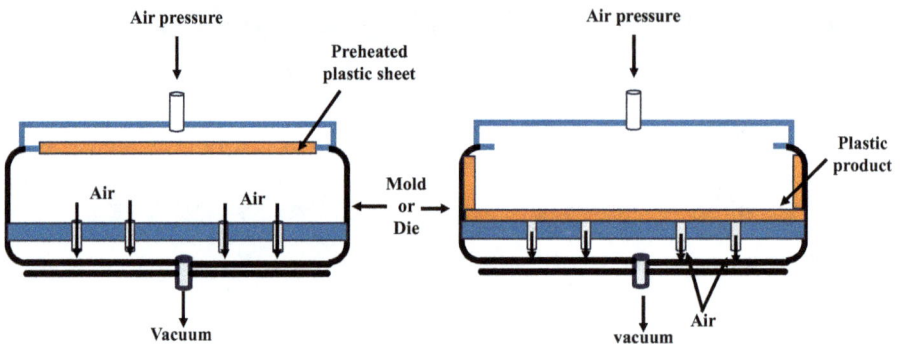

Figure 5.6: Schematic diagram of the polymer thermoforming process.

5.3 Three-Dimensional (3D) Printing

Three-dimensional printing (Figure 5.7), also known as additive manufacturing (AM), is a method of creating three-dimensional items layer by layer using a digital 3D model or computer-aided design (CAD) file. This approach is increasingly being utilized for prototyping and small-scale manufacturing. The procedure starts with producing a 3D model using CAD software. The polymer is then deposited in thin layers using processes such as fused deposition modeling (FDM) and stereolithography, with each layer hardening before the next is added. Following printing, additional finishing operations like curing, cleaning, or support removal may be required to improve the final product.

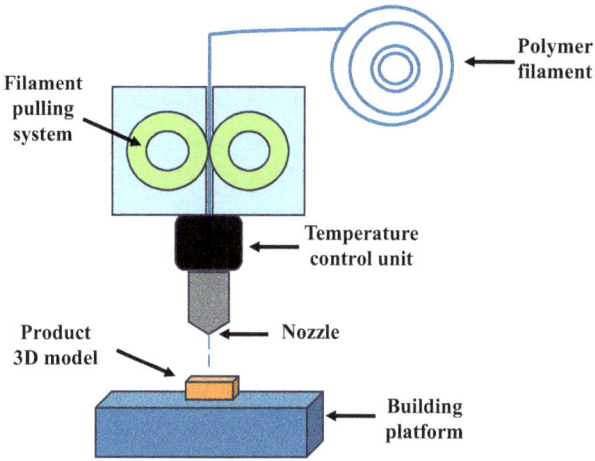

Figure 5.7: Schematic diagram of polymer 3D printing.

5.4 Casting

Casting is one of the cheapest and simplest processes for shaping polymers into the desired solid form. Both thermoplastics and thermosets can be molded by this process. After the pre-preparation of the thermoset or thermoplastic to be cast, the die is loaded with the compound mixture of a thermoset or thermoplastic and then subjected to heating to allow the complete curing of the mixture into a shaped solid form. Once the polymerization or curing is complete, the die is removed and cooled. Casting methods include film casting, die casting, and rotational casting.

5.4.1 Film Casting

Film casting (Figure 5.8) is a manufacturing method that produces thin polymer films or sheets, which are commonly used in packaging, electronics, and medical applications. In this process, the polymer is first dissolved in a solvent to create a homogeneous solution. The solution is then distributed over a flat surface or casting mold. The solvent evaporates, leaving a thin polymer layer. The finished film is then wound into rolls or separated into sheets for further processing and application.

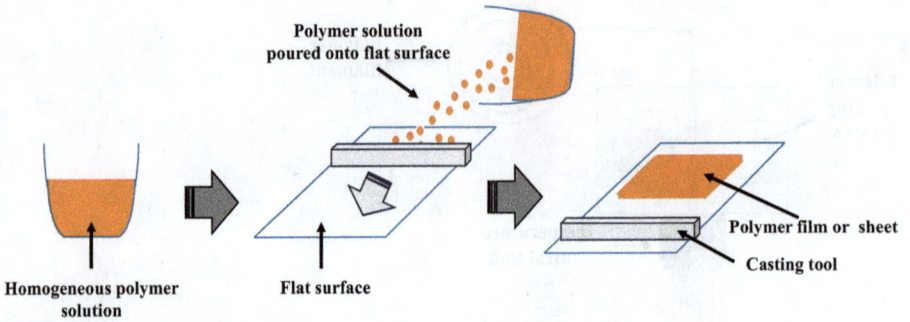

Figure 5.8: Schematic diagram of the polymer film casting process.

5.4.2 Die Casting

Polymer die casting (Figure 5.9) is a manufacturing process where molten or liquid polymer material is injected into a mold cavity under pressure to create complex and detailed parts. Unlike traditional metal die casting, polymer die casting typically involves thermosetting or thermoplastic polymers instead of metals. The material fills the mold, solidifies (through cooling or curing), and is then ejected to form the final product. This technique is commonly used for producing lightweight, durable, and intricate components in industries like automotive and electronics products.

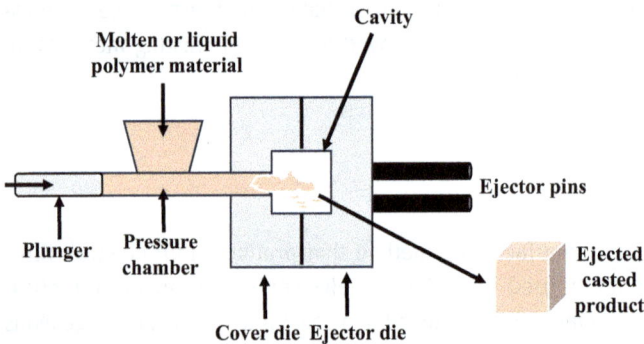

Figure 5.9: Schematic diagram of the polymer die-casting process.

5.4.3 Rotational Casting

In the polymers industry, rotational casting (Figure 5.10) is a technique used to manufacture hollow plastic parts. The process begins by adding powdered thermoplastic into a mold, which is then heated and rotated slowly along two perpendicular axes. As the polymer powder melts, it evenly coats the interior walls of the mold due to the continuous rotational motion. Once the desired thickness is reached, the mold is cooled, solidifying the polymer into its final hollow shape. This method is widely used to produce large, seamless polymer products such as storage tanks, playground equipment, bins, and automotive components. It is particularly valued for its ability to create lightweight, durable parts with uniform wall thickness.

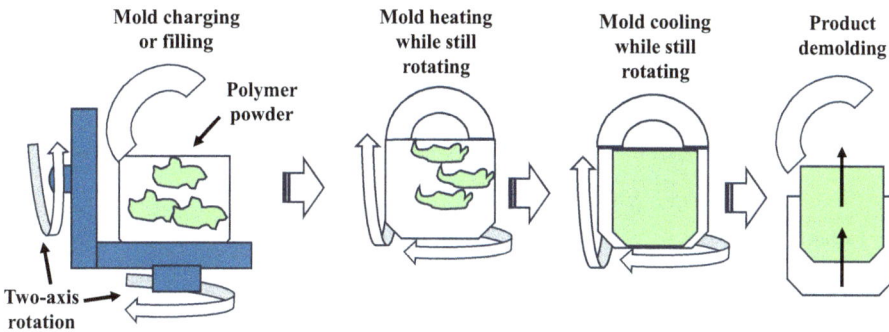

Figure 5.10: Schematic diagram of the polymer rotational casting process.

5.5 Foaming

Foaming (Figure 5.11) is a polymer processing technique that involves injecting gas bubbles into the polymer matrix, resulting in a cellular structure. The process has three stages: cell creation, growth, and stability. Foaming can be accomplished via physical blowing agents (e.g., gas injection) or chemical blowing agents, which decompose and release gas during processing. Depending on the polymer and process used, the resultant foam might be rigid or flexible. Foamed polymers are commonly used in applications such as packaging, insulation, cushioning, and lightweight building materials. This method not only reduces material consumption and weight but also enhances features like thermal insulation, cushioning, and shock absorption.

5.6 Lamination

Lamination (Figure 5.12) in polymers is the process of bonding several layers of materials together to form a composite structure with improved strength, durability, or functional qualities. Typically, adhesives, heat, or pressure are used to bond the layers, which may consist of polymer films, textiles, foams, or other types of materials. During the lamination process, layers are selected based on desired attributes such as strength, flexibility, or barrier performance. Materials are cleaned or pretreated to enhance adherence. The laminated products are subsequently cut, trimmed, or rolled, depending on their intended application. Laminated polymeric materials are utilized in a variety of applications. Selected examples are summarized in Table 5.1.

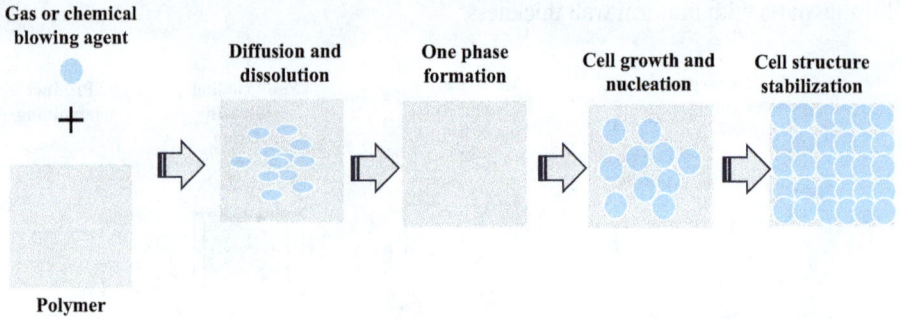

Figure 5.11: Schematic diagram of the foaming process phases.

Table 5.1: Examples of laminated polymer material applications.

Application type	Details
Packaging	Multi-layered films for food and beverage packaging are designed to enhance barrier properties (e.g., moisture or oxygen resistance).
Construction	Laminated panels for flooring, insulation, or decorative surfaces.
Automotive	Laminate safety glass with polymer interlayers.
Textiles	Waterproof and breathable fabrics are made by laminating polymer membranes onto fabrics.

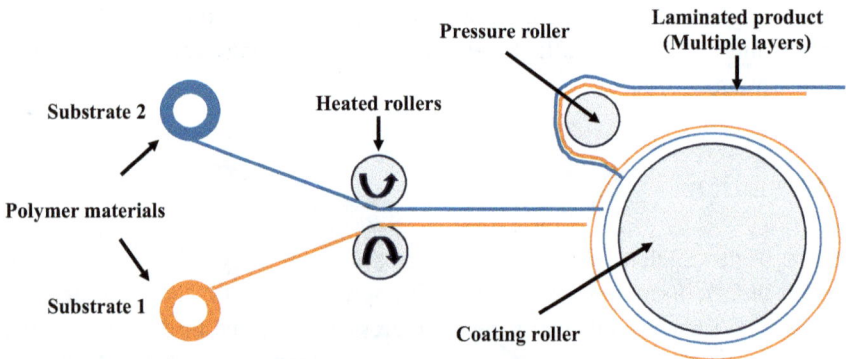

Figure 5.12: Schematic diagram of the lamination process.

5.7 Calendering

Polymer calendering (Figure 5.13) is a manufacturing process that produces polymer films and sheets with precise thickness, surface finish, and mechanical properties. In this process, molten or softened polymer material is passed between a series of heated, counter-rotating rolls, flattening it into a continuous sheet or film. Typically, polymer material in pellet or powder form is melted and softened using heat, with additives such as plasticizers, colorants, and stabilizers incorporated to enhance the final product's properties.

The softened polymer is introduced into the gap between the first set of rolls, where it is compressed and shaped. As the material moves through multiple rollers arranged in sequence, the rolls regulate the sheet's thickness, ensuring uniform density and mechanical properties. These rolls are usually heated to maintain the polymer's pliability throughout the process. Upon exiting the final roll, the polymer sheet is cooled to stabilize its dimensions and may undergo cutting or trimming to achieve the desired width and length. The finished product is then either wound into rolls for flexible films or cut into sheets for further applications.

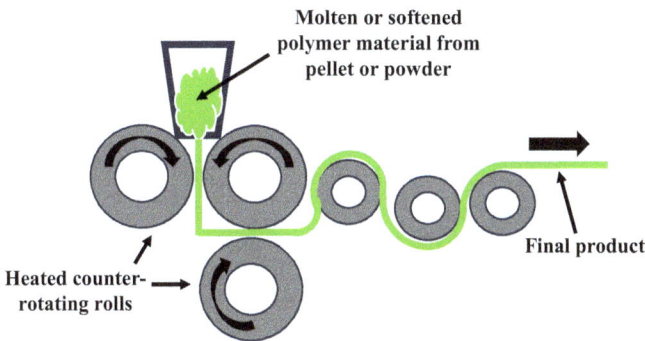

Figure 5.13: Schematic diagram of the calendaring.

5.8 Spinning

Polymer spinning (Figure 5.14) is a process used to create fibers or filaments from polymers, which are later used in textiles, composites, or other applications. It involves the conversion of polymer materials (typically in liquid or molten form) into continuous threads or fibers through a specialized technique called "spinning." This process is commonly applied to synthetic polymers such as nylon, polyester, polypropylene, and acrylic. In the general polymer spinning process, the polymer is either melted or dissolved in a solvent, depending on the spinning method. The polymer solution or melt is then forced through a spinneret, forming thin strands or fibers. The extruded fibers are stretched to align the polymer molecules, enhancing strength and elasticity. The fibers solidify through cooling (melt spinning), evaporation of solvent (dry spinning), or coagulation (wet spinning). There are different types of polymer spinning processes, which are summarized in Table 5.2.

Table 5.2: Types of polymer spinning processes.

Type	Details
Melt spinning	The polymer is melted and extruded through spinnerets (tiny nozzles) into the air, where it solidifies upon cooling.
Solution spinning	It is used for polymers that decompose before melting or are difficult to melt. It has two subtypes: – Dry spinning: The polymer is dissolved in a solvent, extruded through a spinneret, and the solvent is evaporated using hot air. – Wet spinning: The polymer solution is extruded into a coagulation bath containing a non-solvent, causing the polymer to solidify into fibers.
Electrospinning	A polymer solution is subjected to a high-voltage electric field, which draws out ultra-fine fibers and deposits them on a collector.
Gel spinning	A polymer gel is spun into fibers and stretched to enhance molecular alignment.

5.9 Essential Keywords

Blending and compounding Combining polymers with additives, including fillers, plasticizers, stabilizers, and reinforcements, to improve their properties. These methods are vital for tailoring polymer characteristics to specific applications.

Blown film extrusion The method of making thin, flexible films by extruding molten polymer into a tube and inflating it to form thin film sheets. It is widely utilized in the manufacture of packaging films, garbage bags, and agricultural films.

Blow molding A procedure for creating hollow polymer items like bottles and containers. It entails inflating a molten polymer tube inside a mold using air pressure. It is often used to make plastic bottles and large containers.

Calendar rolling (calendering) A method that involves passing a polymer through a sequence of rollers to produce thin sheets or films. It is commonly used to manufacture plastic films, vinyl flooring, and coated textiles.

Casting A manufacturing technique in which a liquid polymer is poured into a mold and left to harden. This process is commonly used to make coatings and films.

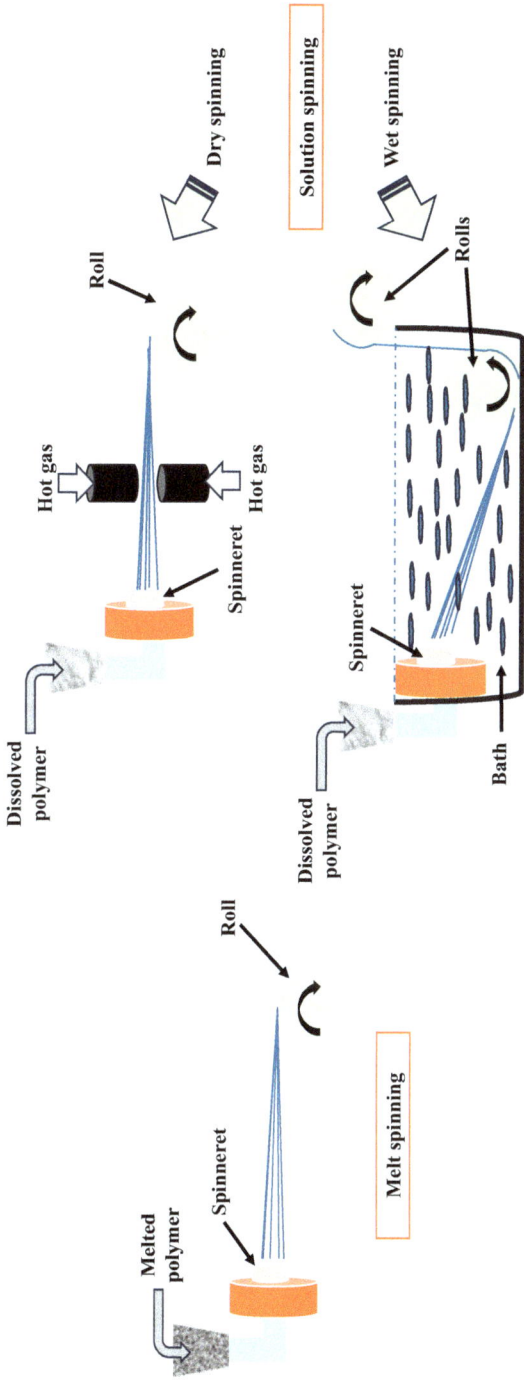

Figure 5.14: Schematic diagram of the types of polymer spinning processes.

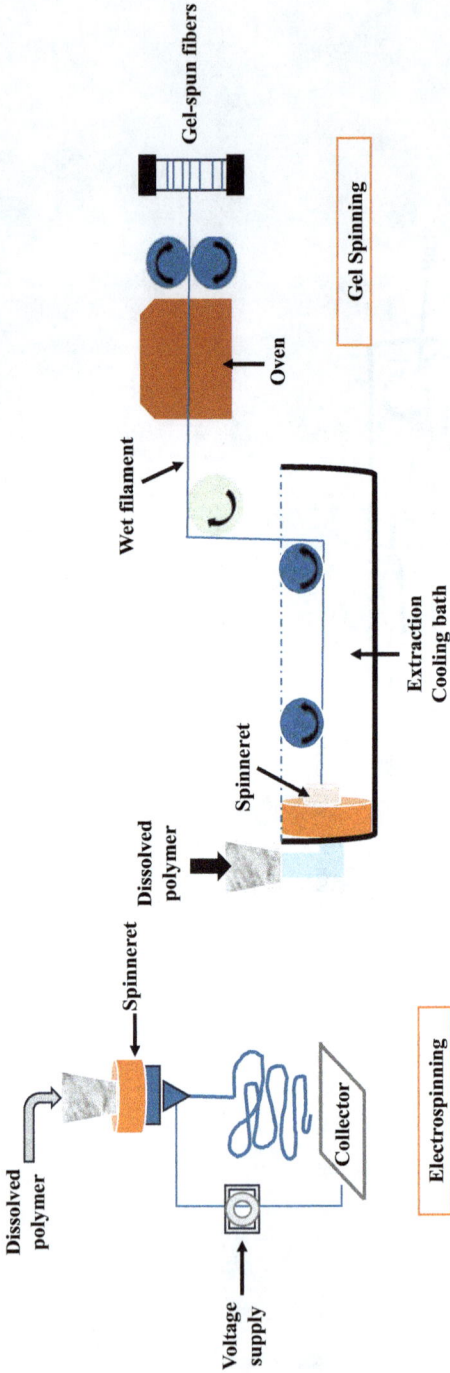

Figure 5.14 (continued)

Co-extrusion Extrusion of two or more different polymers through a multi-layer die to produce a single product with distinct layers. This technique is commonly used in packaging to create barrier films or composite materials.

Compression molding A process in which a polymer material is placed in an open mold cavity and then compressed under heat to form a solid part. This technique is commonly used for large components, such as electrical parts.

Compression thermoforming A process that involves heating a sheet of polymer and placing it into a mold, where it is compressed to form the final part.

Extrusion Involves forcing molten polymer through a die to produce continuous shapes, such as sheets, films, and profiles. The process is highly efficient and is used in producing items like pipes, tubing, and film packaging.

Fiber spinning A process of converting polymer melt or solution into continuous fibers, which are then solidified by cooling or solvent evaporation. This technique is used to produce fibers for textiles, ropes, and filtration applications.

Film blowing A process for producing thin polymer films that are used in packaging, agriculture, and medical applications.

Film and sheet extrusion In this process, polymers are melted and extruded through a flat die to form thin sheets or films. This process is widely used for manufacturing packaging materials, protective films, and automotive components.

Injection molding A widely used polymer processing technique where molten polymer is injected into a mold cavity to form parts. It is ideal for producing complex shapes with high precision and is commonly used in industries such as automotive, medical devices, and consumer goods.

Injection stretch blow molding A combination of injection molding and blow molding. The polymer is first injection-molded into a preform, then stretched and blown into the final shape. This technique is used for producing bottles and containers, particularly in the beverage industry.

Melt spinning A technique in which a polymer is heated to its melting point and then forced through fine nozzles (spinners) to form fibers, which are cooled and solidified.

Rotational molding Involves heating a polymer inside a hollow mold and rotating the mold on two axes to evenly coat the mold surface with the material. This technique is commonly used to create large tanks and playground equipment.

Thermoforming A process in which a polymer sheet is heated until it becomes pliable and is then formed into a specific shape using a mold.

3D printing (additive manufacturing) Involves layer-by-layer deposition of material to build three-dimensional parts directly from digital models.

5.10 Questions and Answers

Questions	Answers
1. What is injection molding, and what are its key applications?	A technique in which a molten polymer is injected under high pressure into a mold is extensively used in several industries, including packaging, consumer goods, and automotive.
2. How does extrusion differ from injection molding in polymer processing?	In extrusion, molten polymer is forced through a die to form continuous shapes like sheets, films, and profiles, while in injection molding, molten polymer is injected into a mold to form discrete parts. Extrusion is typically used for continuous processes, such as making pipes and films, while injection molding is better suited for producing complex, high-precision parts.
3. What is blow molding, and in which industries is it commonly used?	Blow molding is a polymer processing technique used to produce hollow objects by inflating a molten polymer tube inside a mold using air pressure. It is widely used in the production of bottles, containers, and tanks.
4. What is the purpose of thermoforming in polymer processing?	Thermoforming is a process in which a polymer sheet is heated to a pliable state and then formed into a shape using a mold. It is commonly used for producing lightweight parts such as packaging materials, trays, and automotive components. It is also often used for making large parts or thin-walled products like blister packs.
5. How does compression molding work, and for which polymers is it typically used?	Compression molding involves placing a polymer material into an open mold cavity, which is then closed and heated under pressure to form the part. It is typically used for thermosetting polymers and composite materials like rubber and Bakelite. It is ideal for producing large, heavy parts such as automotive bumpers and electrical components.
6. What is the principle behind rotational molding, and what are its advantages?	This method involves heating a polymer inside a hollow mold while rotating it on two axes to distribute the molten material evenly along the interior surface. Its advantages include achieving uniform wall thickness and the ability to create complex shapes without requiring high-pressure equipment.

(continued)

Questions	Answers
7. What are the key features of 3D printing (additive manufacturing) in polymer processing?	3D printing, also known as additive manufacturing, is a method that deposits material layer by layer based on a computer design. This approach allows for the manufacturing of highly customized, sophisticated components while reducing material waste. It is especially useful for applications including quick prototyping, medical implants, and aircraft components.
8. What role do Polymer Blending and Compounding play in processing?	Polymer blending and compounding involve mixing different types of polymers or adding fillers, reinforcements, and additives to achieve desired properties. The process improves characteristics such as strength, flexibility, and thermal stability. This is essential for producing polymers with specific qualities for use in various applications, such as automotive parts and packaging materials.
9. How does fiber spinning contribute to polymer processing?	Fiber spinning is a process used to convert molten polymer or a polymer solution into continuous fibers. This can be achieved through melt spinning, dry spinning, or wet spinning. The process is key to producing synthetic fibers such as nylon, polyester, and polypropylene, which are used in textiles, ropes, filters, and non-woven fabrics.
10. What is the difference between blown film extrusion and cast film extrusion?	Blown film extrusion involves extruding molten polymer into a tube, which is then inflated to create a thin, flexible film. In contrast, cast film extrusion involves extruding molten polymer onto a flat surface or casting die, resulting in a thin film with a more uniform thickness and smoother surface.
11. What are the four types of polymer spinning processes?	– Melt spinning – Solution spinning – Electrospinning – Gel Spinning
12. What is rotational molding used to produce?	It produces hollow plastic products such as large tanks, containers, and playground equipment.
13. What does calendaring involve?	This process involves passing molten or softened polymer material between a series of heated, counter-rotating rolls to flatten it into a continuous sheet or film.

(continued)

Questions	Answers
14. Define foaming.	Foaming in polymers is a process used to create lightweight materials by introducing gas bubbles into the polymer matrix, forming a cellular structure.
15. What does co-extrusion stand for?	The simultaneous extrusion of two or more different polymers through a multi-layer die produces a single product with distinct layers.

Chapter 6
Cutting-Edge Polymers

Cutting-edge polymers represent a significant advancement in material science, offering new solutions to global challenges and expanding the possibilities for new polymer materials that provide specialized functions, higher performance, and reduced environmental impact. These polymers are novel and advanced materials that exhibit unique properties, functionalities, and applications not typically found in conventional polymers. They are often designed or developed to meet the evolving needs of various industries, such as electronics, healthcare, environmental sustainability, energy, and biotechnology. Examples include biodegradable polymers, conductive polymers, polymers for drug delivery, and polymers with advanced mechanical, thermal, and optical properties. Additional polymer types can be categorized under this group, and some of them are briefly considered and discussed in the paragraphs to follow.

6.1 Biodegradable Polymers

Biodegradable polymers are a category of polymers that break down into environmentally friendly byproducts under natural conditions. They play a vital role in minimizing plastic waste and are widely utilized in packaging, medical devices, and agricultural applications. Table 6.1 provides an overview of selected examples.

Table 6.1: Examples of biodegradable polymers.

Polymer name	Usage	Advantages	Challenges
Polylactic acid (PLA)	It is used for biodegradable packaging, disposable utensils, and medical devices.	Environmentally friendly, reducing long-term pollution, degrading into non-toxic byproducts, and suitable for applications where traditional plastics are not ideal.	They are higher in cost compared to conventional plastics and have limited mechanical properties, as well as processing difficulties.
Polyhydroxyalkanoates (PHA)	These are used in packaging, agricultural films, and medical applications.		
Polycaprolactone (PCL)	Used in drug delivery systems, tissue engineering, and biodegradable plastics.		

https://doi.org/10.1515/9783111585734-006

6.2 Conductive (Conducting) Polymers

Conductive polymers are organic materials capable of conducting electricity, either intrinsically or upon doping with specific agents such as 2-naphthalene sulfonic acid (NSA), arsenic pentafluoride (AsF_5), and methane sulfonic acid (CH_3SO_3H). Notable examples include polypyrrole, commonly used in antistatic coatings; polyaniline, which provides corrosion protection; and polythiophene, a key component in solar cells and transistors. Their chemical structures are illustrated in Figure 6.1.

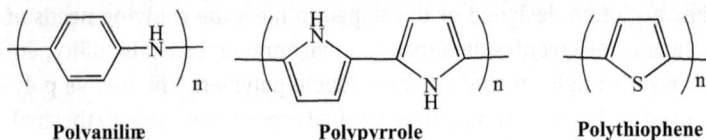

Polyaniline   Polypyrrole   Polythiophene

Figure 6.1: Selected examples of conductive polymers.

6.3 Biomedical Polymers

Polymers for drug delivery (biomedical polymers) are designed to control the release of drugs in a targeted and sustained manner. These polymers are often biocompatible, making them suitable for use in the human body. They can be used for the encapsulation of drugs, ensuring that the drug is delivered to the right place at the right time, which improves efficacy and reduces side effects. Their use extends from oral and mucoadhesive drug delivery systems to colon drug delivery systems. Among these are polycaprolactone (PCL), polylactic acid (PLA), polylactide-co-glycolic acid (PLGA), and Eudragit polymers (copolymers derived from esters of acrylic and methacrylic acid). Several synthetic polymers have also been used as polymeric implant materials. Selected examples include polydimethylsiloxane (PDMS), polyethylene (PE), polyetheretherketone (PEEK), and polyurethane (PU). Other polymers are used as carriers of bioactive substances (polymeric supports). Examples include polystyrene (PS) and ethylene-maleic anhydride (EMA) copolymers.

6.4 Polymeric Supports for Solid-Phase Synthesis in Oligonucleotides and Polypeptides

Organic synthesis on the solid phase consists of a sequence in which the starting material is covalently bound to a swollen, insoluble polymeric support through an anchor and/or a linker. During all synthetic steps, the compound under construction remains linked to the support. The most obvious advantage of this method is the possibility of using an excess of reagents to complete the reaction, as the removal of unreacted reagents can be achieved through simple filtration. The most commonly used polymeric support is a copolymer of styrene/divinylbenzene, functionalized with various reactive groups such as chloromethyl (Merrifield resin) and various spacers such as ethylene glycol (Wang-type resin). For solid-phase synthesis of oligonucleotides, many types of solid supports have been used, but controlled pore glass (CPG) and polystyrene have proven to be the most useful.

6.5 High-Performance Polymers

High-performance polymers (HPPs) have superior mechanical, thermal, and chemical properties compared to ordinary polymers. These materials, known for their exceptional strength, chemical resistance, and ability to withstand higher temperatures, are widely employed in aerospace, automotive, and medical implant applications. Figure 6.2 depicts examples such as polyphenylene sulfide (PPS), polyetherimide (PEI), polyethersulfone (PES), polyetheretherketone (PEEK), and polyimide (PI).

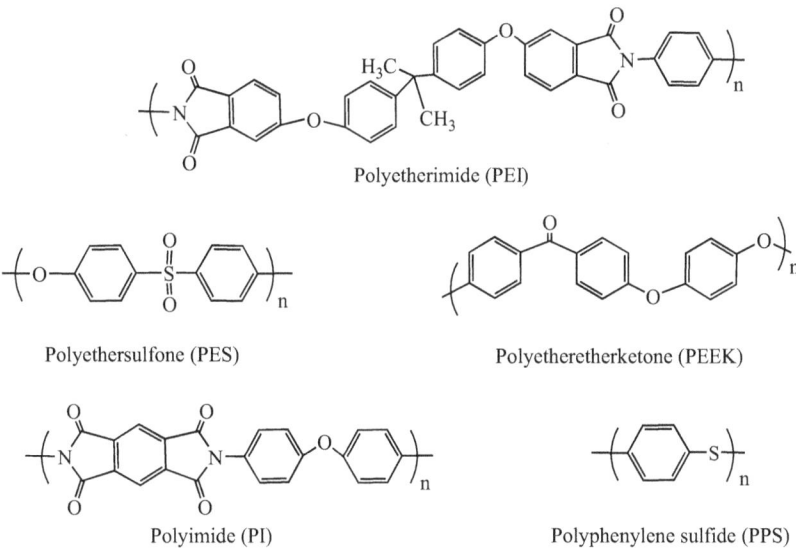

Polyetherimide (PEI)

Polyethersulfone (PES)

Polyetheretherketone (PEEK)

Polyimide (PI)

Polyphenylene sulfide (PPS)

Figure 6.2: Selected examples of high-performance polymers.

6.6 Smart Polymers

Smart polymers (or stimuli-responsive polymers) are materials that change their physical or chemical properties in response to external stimuli such as temperature, pH, light, or electric fields. These polymers can be used in sensors, actuators, and drug delivery systems that respond dynamically to environmental changes. Biological applications of this technology currently under development span diverse areas, including bioseparation, drug delivery, reusable enzymatic catalysts, molecular switches, biosensors, regulated protein folding, microfluidics, and gene therapy. Examples include thermoresponsive polymers (temperature-responsive polymers) such as poly(N-isopropylacrylamide) (PNIPAM), which exhibit a dramatic change in solubility or phase transition when the temperature changes. These are commonly used in drug delivery and tissue engineering. pH-responsive polymers, which change their structure or solubility in response to pH changes, are used for controlled drug release in the stomach or intestines, and photoresponsive polymers, which undergo structural changes when exposed to light, are useful in applications such as drug delivery and photonic devices.

6.7 Environmentally Sustainable Polymers

Polymers supporting environmental sustainability are quickly becoming viable alternatives to traditional plastics. These polymers are both biodegradable and derived from renewable resources. The aim of their use is to reduce the environmental impact of polymer waste while still delivering the necessary performance for modern applications. Examples include bio-based polymers made from renewable resources such as plant sugars, oils, and starches. Another example is recycling-enhanced polymers, which are designed for easy recyclability or to incorporate recycled materials, such as recovered PET (rPET) used in sustainable packaging.

6.8 Water-Soluble Polymers

Water-soluble polymers are a diverse and versatile class of materials with applications across many industries, including pharmaceuticals, food, agriculture, and wastewater treatment. Their ability to dissolve in water and form gels, solutions, or films, coupled with their biocompatibility and biodegradability, makes them invaluable for a wide range of functional applications. Water-soluble polymers typically have hydrophilic (water-attracting) functional groups in their molecular structure, such as hydroxyl (-OH), carboxyl (-COOH), or amine ($-NH_2$) groups. These groups interact with water molecules, allowing the polymer chains to swell or dissolve in water. The solubility of these polymers depends on factors such as the degree of polymerization, the presence of ionic or polar groups, and environmental conditions such as temperature and pH. Different types of these polymers are summarized in Table 6.2.

Table 6.2: Different types of water-soluble polymers.

Type	Application
Polyethylene glycol (PEG)	It is used in drug delivery systems as a solubilizing agent, for making ointments and creams, and as a food and beverage ingredient.
Polyvinyl alcohol (PVA)	It is used in biodegradable packaging materials, tablet coatings, and as a binder in pharmaceutical formulations. It is also used to improve soil structure and water retention in agriculture, thereby reducing erosion and increasing crop growth.
Carboxymethyl cellulose (CMC)	It is used as a thickener, stabilizer, and emulsifier in food items, as well as a disintegrant in tablets and a suspending agent in oral suspensions.
Polyacrylic acid (PAA)	Used in controlled drug delivery systems, as a stabilizing agent for suspensions, and in water treatment during flocculation processes to remove impurities from water.
Polyvinylpyrrolidone (PVP)	Used as a binder, solubilizer, and stabilizer in drug formulations and oral solutions.
Polyacrylamide (PAM or pAAM)	Used in a wide range of cosmetic products, such as moisturizers, lotions, creams, and self-tanning products, it also finds applications in pulp and paper production, agriculture, food processing, mining, and as a flocculant in wastewater treatment.
N-(2-Hydroxypropyl) methacrylamide (HPMA)	Very successfully used for passive drug-targeting purposes and as a carrier for low-molecular-weight drugs.
Divinyl ether-maleic anhydride (DIVEMA)	Has shown antitumor activity against various types of tumors.
Xanthan gum	A free-flowing powder that dissolves in both hot and cold water, forming viscous solutions even at low concentrations. Its industrial significance lies in its ability to regulate the rheology of water-based systems. It serves as a highly effective thickener and stabilizer, producing significantly more viscous solutions than other polysaccharides at comparable concentrations.
Pectin	Pectin has been used in the pharmaceutical industry for a wide range of applications. Pure and standardized pectin has been used as a binding agent in tablets. High Methoxy (HM) pectin is used as a monolithic bioerodible system and in the preparation of directly compressible tablets along with HPMC. Low Methoxy (LM) pectin has been used to prepare beads by the ionotropic gelation technique and for sustained-release drug delivery using calcium pectinate gel beads.

6.9 Electroluminescent Polymers

Electroluminescent polymers (ELPs) are a class of conjugated polymers that emit light in response to an applied electric current or voltage. These polymers have a conjugated system of alternating single and double bonds in their molecular structure, which allows them to conduct electricity and emit light when excited by an electric field. Due to their unique properties, electroluminescent polymers are used in a variety of applications, including displays, lighting, and advanced electronics. Electroluminescent polymers possess several unique properties that make them valuable in various technological applications, as summarized in Table 6.3.

Table 6.3: Technological applications of electroluminescent polymers.

Applications	Details
Conjugation	The polymer chain contains conjugated π-electrons, allowing for the movement of electrons, and light emission occurs when these electrons recombine.
Light emission	As the electric current passes through the polymer, electrons are injected into the polymer and recombine with holes. This recombination releases energy in the form of light.
Tunable emission	The wavelength (color) of the emitted light can be tuned by adjusting the chemical structure of the polymer. Different functional groups and monomers can be incorporated to produce polymers that emit various colors, ranging from blue and green to red and white.
Solution processability	Electroluminescent polymers can be processed from solution, meaning they can be applied as thin films using techniques such as spin coating, inkjet printing, or spray deposition. This makes them suitable for large-area applications and flexible substrates.
Flexibility and lightweight	Many electroluminescent polymers are inherently flexible and lightweight, which makes them suitable for use in flexible electronics and displays.
Low voltage operation	ELPs can be designed to operate at relatively low voltages, making them efficient and easier to integrate into low-power devices.

There are various types of electroluminescent materials, each with distinct structures, properties, and specific applications. Table 6.4 presents selected examples, and their structures are illustrated in Figure 6.3.

Electroluminescent polymers offer several advantages, including solution processability, flexibility, customizable emission, and low-voltage operation. However, despite these benefits, certain challenges remain. One major issue is their limited stability, especially under prolonged use or exposure to oxygen and moisture, which can lead to reduced brightness and shorter device lifetimes. Additionally, their light emission efficiency may still lag behind that of inorganic LEDs in some applications. Pro-

Table 6.4: Examples of electroluminescent polymers.

Name	Structure details	Properties	Specific applications
Poly (p-phenylene vinylene) (PPV)	It consists of a backbone of alternating phenylene (benzene) rings and vinylene (CH = CH) units. Derivatives of PPV, such as MEH-PPV (poly [2-methoxy-5-(2'-ethylhexyloxy)-1,4-phenylene vinylene]), are more soluble and can be processed more easily.	These polymers typically emit light in the orange-to-red region of the spectrum. They are often used in OLED (organic light-emitting diode) displays and light-emitting diodes.	Used in organic light-emitting diodes (OLEDs), displays, and lighting applications.
Polyfluorenes (PFs)	A polymer with a backbone of fluorene units. Fluorene-based polymers are known for their high efficiency and excellent stability.	These polymers emit blue light and are used in applications requiring high brightness and color purity.	Blue OLEDs, display technology, and lighting.
Polythiophenes (PTs)	Polythiophene polymers consist of thiophene rings as the monomer units. Thiophene-based polymers have conjugated systems that can be modified to achieve different properties.	Polythiophene derivatives can emit a range of colors, from red to green. They are known for their excellent conductivity and stability.	OLEDs, solar cells, sensors, and transistors.
Poly(para-phenylene) (PPP)	PPP is a conjugated polymer with a rigid and planar backbone structure.	It can emit bright green light and is used in various display technologies.	Displays, lighting, and optoelectronics.
Poly(phenylene ethynylene) (PPE)	PPE is a conjugated polymer that features phenylene units connected by triple bonds, creating a linear structure with a high degree of conjugation.	PPE polymers can emit light in various colors and are characterized by their excellent luminescence efficiency.	Light-emitting devices and sensors.

cessability and scalability also pose challenges. While solution processing is beneficial, ensuring uniformity and high quality in large-scale production remains a hurdle, particularly for high-performance applications.

6.10 Polymers Microgels

Polymer microgels are crosslinked polymer particles within the colloidal size range that can expand and contract in response to environmental changes. Their distinctive

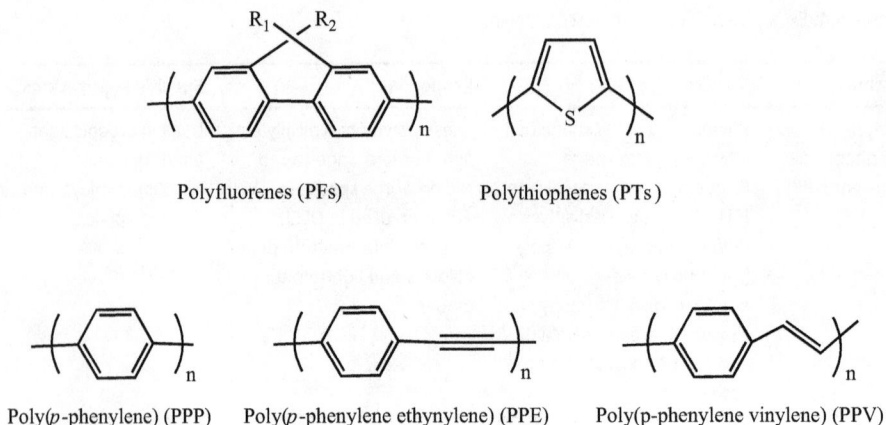

Polyfluorenes (PFs)   Polythiophenes (PTs)

Poly(*p*-phenylene) (PPP)   Poly(*p*-phenylene ethynylene) (PPE)   Poly(p-phenylene vinylene) (PPV)

Figure 6.3: Selected examples of electroluminescent polymers.

properties make them valuable for applications in drug delivery, biotechnology, and responsive materials. As crosslinked polymer networks, they can retain water while preserving their structural integrity. They are often categorized based on their cross-linking mechanism, charge, or interaction behavior. The classification into Type A and Type B microgels typically refers to their surface charge and interaction with solvents or biomolecules. The key differences between Type A and Type B microgels are summarized in Table 6.5.

Table 6.5: The key differences between Type A and Type B microgels.

Microgel type	Charge	Common polymers	Interactions	Applications
Type A microgels	Positively charged (cationic)	Often made from cationic polymers such as poly(ethyleneimine) (PEI) or poly (diallyldimethylammonium chloride) (PDADMAC).	Interact strongly with negatively charged biomolecules, proteins, or drug molecules.	Used in gene delivery, antibacterial coatings, and protein immobilization.
Type B microgels	Negatively charged (anionic)	Composed of polymers such as poly(acrylic acid) (PAA), alginate, or carboxymethyl cellulose (CMC).	Bind effectively with positively charged drugs, proteins, or metal ions.	Commonly used in drug delivery, wastewater treatment (for metal ion removal), and bioengineering.

Microgels can also be further classified based on the incorporation of various poly-meric or non-polymeric materials into their structure to improve both their adsorp-tive properties and their ability to regenerate after adsorbing pollutants. This further classification is summarized in Table 6.6.

Table 6.6: Microgel classification based on the incorporation of various polymeric or non-polymeric materials.

Class	Details
Simple microgels	Consists solely of a monomer, either alone or in combination with comonomers. An example is microgels based on poly(N-isopropylacrylamide).
Core-shell microgels	Core-shell microgels feature a distinct morphology consisting of two regions: a core and a shell. Both the core and shell can be composed of either organic or inorganic materials. A notable example is an inorganic core encapsulated by an organic polymer shell, where the core is made of inorganic substances such as SiO_2, Fe_2O_3, or Fe_3O_4, while the shell consists of crosslinked organic polymers.
Homogenous inorganic-organic microgel composite	In homogeneous inorganic–organic composite systems, the inorganic material (SiO_2, Fe_2O_3, or Fe_3O_4) is evenly dispersed within the crosslinked network of organic polymers.
Fe_3O_4-crosslinked organic polymer microgels	In Fe_3O_4-crosslinked organic polymer microgel systems, the organic monomer, either alone or with a comonomer, is crosslinked with Fe_3O_4 as the inorganic component.

6.11 Essential Keywords

Biodegradable polymers Molecules designed to break down in natural environments through microbial activity. These polymers, such as polylactic acid and polyhydrox-yalkanoates, are increasingly used in packaging and medical applications to reduce environmental impact.

Biomedical polymers Molecules used for drug delivery and are designed to control the release of drugs in a targeted and sustained manner.

Biopolymers Molecules derived from renewable natural resources, such as plants or microorganisms. Examples include cellulose, chitosan, and starch-based polymers, which have applications in packaging, agriculture, and healthcare.

Conductive polymers Organic materials that can conduct electricity. They are used in applications such as flexible electronics, sensors, and antistatic coatings.

Electroluminescent polymers Are a class of conjugated polymers that emit light in response to an applied electric current or voltage.

Green polymers Are sustainable materials that minimize environmental impact throughout their lifecycle. They include biodegradable, recyclable, and bio-based polymers that contribute to circular economy initiatives and environmentally friendly packaging.

High-performance polymers Molecules with superior mechanical, thermal, and chemical properties compared to conventional polymers.

Polymer microgels Are crosslinked polymer particles within the colloidal size range that can expand and contract in response to environmental changes.

Photoresponsive polymers Molecules that undergo structural changes when exposed to light.

pH-responsive polymers Molecules that change their structure or solubility in response to pH changes and are used for controlled drug release in the stomach or intestines.

Smart polymers (or stimuli-responsive polymers) Are materials that change their physical or chemical properties in response to external stimuli such as temperature, pH, light, or electric fields.

Superabsorbent polymers Molecules that absorb and retain large amounts of liquid relative to their weight. These polymers are commonly used in diapers, feminine hygiene products, and medical wound dressings.

Sustainable polymers Molecules derived from renewable sources, such as lignin and bio-based polyesters.

Thermoresponsive polymers (temperature-responsive polymers) Macromolecules that exhibit a dramatic change in solubility or phase transition when the temperature changes.

Water-soluble polymers Molecules that dissolve in water and form gels, solutions, or films.

6.12 Questions and Answers

Questions	Answers
1. What are cutting-edge polymers?	Cutting-edge polymers are advanced polymeric materials designed with innovative properties for high-performance applications in fields such as medicine, electronics, and sustainability.
2. What are biodegradable polymers, and why are they important?	Biodegradable polymers are materials designed to break down over time through natural processes, such as microbial action. They are important because they offer an alternative to conventional plastics, reducing environmental pollution, especially in packaging and disposable products.
3. Give two examples of biodegradable polymers.	Polylactic acid (PLA) and polyhydroxyalkanoates (PHA).
4. What are conductive polymers, and what are their applications?	Conductive polymers are organic polymers that can conduct electricity, unlike typical insulating polymers. They are used in flexible electronics, batteries, sensors, and antistatic materials.
5. What are high-performance polymers, and where are they used?	High-performance polymers are materials that exhibit excellent mechanical, thermal, and chemical resistance properties, even at extreme temperatures.
6. What are smart polymers, and how do they work?	Polymers that can respond to environmental stimuli, such as temperature, pH, or light, are ideal for advanced applications.
7. What is the significance of biopolymers in emerging technologies?	They are biodegradable materials derived from renewable resources and can be used in applications such as packaging, agriculture, and medicine, contributing to eco-friendly and sustainable technology development.
8. Give examples of high-performance polymers.	Polyetheretherketone (PEEK) and polyimide (PI).
9. What are the uses of smart polymers?	Smart polymers are used in drug delivery systems, biosensors, and self-healing materials.
10. What are superabsorbent polymers (SAPs), and where are they used?	Polymers have the ability to absorb and retain multiple times their weight in water or other liquids. They are commonly used in products such as diapers, sanitary napkins, and medical wound dressings due to their high absorbency and moisture retention properties.

(continued)

Questions	Answers
11. What are self-healing polymers, and how are they used?	Self-healing polymers contain microcapsules or reversible bonds that allow the material to restore its original properties. They are used in coatings, automotive parts, and electronics to improve durability and lifespan.
12. What are the environmental benefits of using green polymers?	Green polymers help reduce dependence on fossil fuels and lower the environmental impact of polymer products.
13. What are the potential applications of polymers in energy storage?	Polymers in energy storage play a key role in the development of lightweight, flexible, and high-capacity batteries and supercapacitors. Polymers with high ionic conductivity are utilized in the design of flexible batteries for applications such as wearable electronics, electric vehicles, and energy storage systems that demand efficient and lightweight materials.
14. What role do polymers play in 3D printing?	Polymers such as PLA and ABS are widely used in 3D printing for creating lightweight, durable, and customizable structures.
15. How do conductive polymers work?	Conductive polymers have conjugated electron systems that allow them to conduct electricity, making them useful in flexible electronics and sensors.

Chapter 7
Polymer Product Design and Applications

Product design and polymer processing technologies are closely interconnected, as advancements in polymer science and technology directly influence the development of innovative products. Polymer processing technologies provide the foundation by exploring material properties, synthesis, manufacturing methods, and processing techniques, ensuring that polymers meet specific functional and performance requirements. Product design, however, applies this knowledge to create practical and efficient solutions tailored to industry needs, balancing esthetics, durability, sustainability, and cost. The synergy between these fields enables the creation of high-performance materials for applications in industries such as healthcare, automotive, aerospace, and packaging. This, in turn, plays a transformative role in a wide array of industries, from packaging and electronics to medical devices and textiles. With ongoing innovations in sustainable practices, biodegradable materials, and advanced applications like organic electronics, polymers continue to drive technological progress while offering more environmentally responsible production opportunities. Polymer product design involves selecting appropriate polymer materials and technological processing techniques to create products with desired properties, such as durability, flexibility, chemical resistance, or thermal stability. Engineers and designers consider factors like mechanical performance, environmental impact, cost, and manufacturing feasibility.

7.1 Product Design

Polymer product design is a critical discipline in the field of industrial design and engineering, focusing on developing products made from polymer materials. It integrates knowledge of materials science, mechanical engineering, and manufacturing processes to create functional, durable, and cost-effective products. Polymers are versatile materials with a range of properties that make them suitable for a wide array of applications, from consumer goods and packaging to automotive and medical devices. Polymer product design involves careful consideration of several factors, including material selection, processing methods, functionality, sustainability, and cost efficiency.

Product design using polymers involves understanding the characteristics and behavior of different types of polymers and applying this knowledge to develop a product that fulfills specific requirements. Polymers are organic compounds made up of long chains of molecules, and their physical properties can be manipulated by changing the molecular structure or by adding various additives. The broad array of polymer types offers designers numerous options for tailoring materials to meet the needs of their products. The significance of polymer product design is highlighted for various reasons, as summarized in Figure 7.1.

https://doi.org/10.1515/9783111585734-007

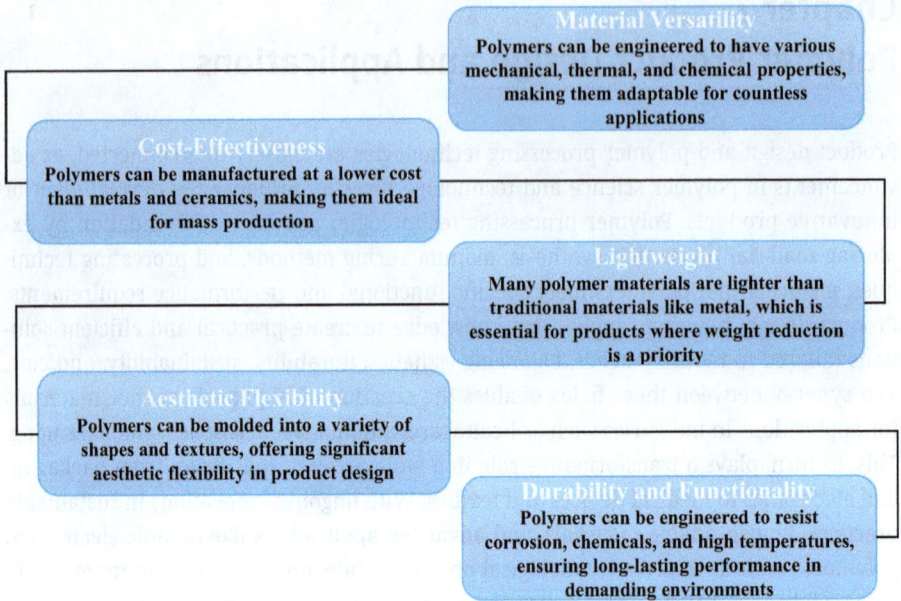

Figure 7.1: Significance of polymer product design.

7.1.1 Key Principles of Polymer Product Design

The process of polymer product design is driven by several guiding principles that focus on balancing functionality, manufacturability, and cost. These principles form the foundation upon which successful designs are built.

7.1.1.1 Material Selection
One of the first and most important steps in polymer product design is selecting the appropriate polymer material. The choice of material affects the product's performance, manufacturing process, and cost. Designers must consider the following factors:

- **Mechanical Properties:** Depending on the product's application, properties such as tensile strength, stiffness, impact resistance, and elasticity must be considered. For example, an automotive bumper needs a polymer that can withstand high impact without breaking, while a flexible seal requires a material with high elasticity.
- **Thermal Properties:** Polymers display diverse levels of heat resistance. For example, polytetrafluoroethylene (PTFE) is highly stable at elevated temperatures, while polyethylene (PE) is more susceptible to thermal degradation.
- **Chemical Resistance:** Certain polymers are highly resistant to chemicals, while others are more vulnerable. This is critical for products exposed to harsh environments, such as automotive fuel tanks or chemical storage containers.

- **Processing Properties:** The ease with which a polymer can be molded, extruded, or shaped significantly influences material selection and design. Common polymers, such as polyethylene and polypropylene, are highly processable, while advanced materials, such as polyetheretherketone (PEEK), are more challenging to manufacture due to their specialized processing requirements.
- **Cost:** The cost of raw materials is a significant factor. Designers must balance the desired material properties with the budget available for manufacturing.

7.1.1.2 Design for Manufacturability

Design for Manufacturability (DFM) is a crucial consideration in polymer product design. This principle emphasizes designing products in a way that makes them easy, efficient, and cost-effective to manufacture. Some factors to consider in DFM include:

- **Part Complexity:** Overly complex parts with intricate features may be difficult or expensive to manufacture. Simplifying the design to reduce the number of features can significantly lower production costs.
- **Material Waste:** Efficient use of materials minimizes waste, lowers costs, and improves sustainability.
- **Assembly:** A well-designed product should minimize the number of components, thereby reducing the need for extensive assembly.
- **Mold Design:** For injection molding and other processes, mold design is critical. The part must be designed in such a way that it can be easily ejected from the mold without damage, and the mold itself must be cost-effective and durable.

7.1.1.3 Functionality and Performance

A core objective in product design is to ensure the product performs its intended function efficiently and reliably. This requires a careful evaluation of the physical forces involved and the selection of a polymer capable of withstanding those conditions. Key functional considerations include:

- **Load-Bearing Capacity:** The polymer material must withstand mechanical stress without deforming or failing.
- **Flexibility:** For components requiring pliability, such as hoses, seals, or gaskets, elastomers or flexible thermoplastics are preferred for their ability to bend without losing performance.
- **Impact Resistance:** Products that are exposed to impact, such as automotive parts or consumer electronics, require polymers with high impact resistance.
- **Surface Finish:** The surface texture of the polymer product can influence its performance and appearance. Smooth surfaces may be important for products such as food packaging, while textured surfaces may be required for better grip on handles or tools.

7.1.1.4 Sustainability

As sustainability gains prominence in manufacturing, polymer product design has adapted to incorporate environmentally responsible practices. Key approaches include:

- **Recyclability:** Selecting materials that can be efficiently recycled helps minimize the environmental footprint of polymer-based products.
- **Biodegradable Polymers:** For applications where long-term durability is not required, biodegradable polymers provide a sustainable alternative to conventional plastics, reducing waste and environmental impact.
- **Energy Efficiency:** Enhancing production processes, such as optimizing molding conditions, reduces energy consumption and minimizes the product's overall carbon footprint.
- **Life Cycle Assessment (LCA):** Assessing a product's environmental impact throughout its entire life cycle, from raw material extraction to disposal, helps ensure that sustainability is considered at every stage of design and manufacturing.

7.1.1.5 Aesthetics and User Interaction

The design should consider both the appearance and the usability of the product. Esthetic considerations include color, surface texture, transparency, and overall appearance. Additionally, the user experience (UX) plays a significant role in the design of products such as consumer electronics, medical devices, and kitchenware. For example:

- **Color and Texture:** Polymers can be easily colored and textured. Designers may choose specific colors for branding, functionality, or user preferences.
- **Ergonomics:** Polymers can be designed for ergonomic use, ensuring comfort and ease of use. Products such as handheld tools, medical instruments, and consumer electronics are often designed with ergonomics in mind.
- **User Safety:** Polymers must be selected or treated to ensure they are safe for their intended use. This is particularly important in products such as medical devices, toys, and food packaging.

7.2 Key Industrial Sectors and Applications

7.2.1 Automotive Industry

Polymers are essential in the automotive industry (Figure 7.2) due to their lightweight properties, which improve fuel efficiency and overall vehicle performance. They are extensively used in both interior and exterior components, including dashboards, bumpers, door panels, lighting, and seating.

Materials such as polypropylene (PP) and polyethylene (PE) provide excellent durability and flexibility, while polyurethane (PU) enhances insulation, cushioning, and noise reduction. The incorporation of polymer-based components reduces vehicle weight, contributing to greater energy efficiency and lower emissions.

Additionally, the demand for high-performance thermoplastic polymers, such as polycarbonate (PC) and acrylonitrile-butadiene-styrene (ABS), continues to grow for applications requiring exceptional impact resistance and esthetic appeal.

Figure 7.2: Automotive polymers for a wide range of components (image credit: https://www.elastomer. kuraray.com/applications/automotive-polymers/).

7.2.2 Aerospace Industry

Polymers are also used in aerospace (Figure 7.3) for lightweight and high-strength materials required in aircraft construction. Components like cabin interiors, insulation, fuel tanks, and certain structural elements rely on polymers such as carbon fiber-reinforced composites and thermoplastics. Polymers contribute to reducing the overall weight of the aircraft, which improves fuel efficiency and reduces operational costs. They are also utilized in thermal insulation, electrical insulation, and anti-vibration systems. Advanced polymers, such as thermosetting composites, are employed in critical parts due to their high thermal stability and resistance to harsh conditions, ensuring safety and performance in demanding aerospace applications.

7.2.3 Electronics Industry

In electronics (Figure 7.4), polymers are used for insulation, casings, displays, and wires. Conductive and semi-conductive polymers are being increasingly integrated into electronic devices, sensors, and touchscreens. Polymers in electronics reduce weight and cost, offer electrical insulation, and allow for flexible or lightweight devices. They also enhance device durability and ease of manufacturing. Conductive polymers are being explored for use in organic light-emitting diodes (OLEDs), flexible electronics, and organic photovoltaics, paving the way for new forms of technology such as foldable or stretchable screens. The use of conductive polymers in organic semiconductors is enabling the development of printable electronics, including sensors, transistors, and flexible displays. Polymers are also used in batteries and energy storage devices, such as organic solar cells.

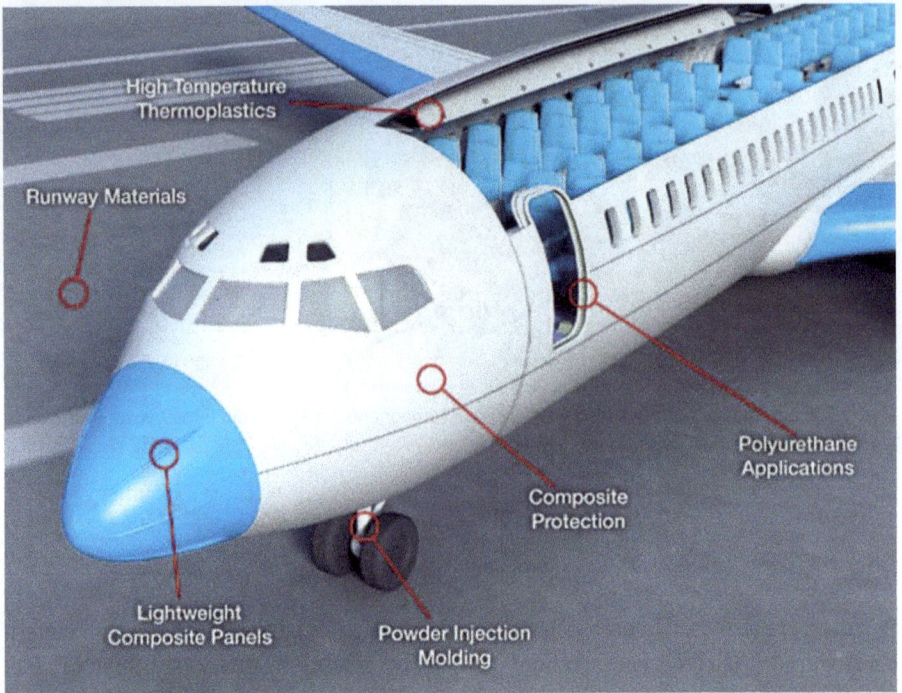

Figure 7.3: Adhesives and coatings in the aerospace industry (image credit: https://nationalpolymer.com/aerospace/).

7.2.4 Medical and Healthcare

Polymers are widely utilized in the medical field (Figure 7.5) due to their versatility, biocompatibility, and adaptability for specific applications. They are essential in the manufacturing of medical devices such as catheters, syringes, implants, prosthetics, and drug delivery systems. Biocompatible and easily sterilizable polymers like polyethylene (PE), polypropylene (PP), and polyvinyl chloride (PVC) are commonly used in medical devices due to their safety and durability. Additionally, biodegradable polymers such as polylactic acid (PLA) and polycaprolactone (PCL) play a crucial role in implants and drug delivery systems, as they gradually break down and are absorbed by the body over time. Polymers are also used in controlled or sustained drug release systems, allowing drugs to be delivered over a prolonged period without the need for

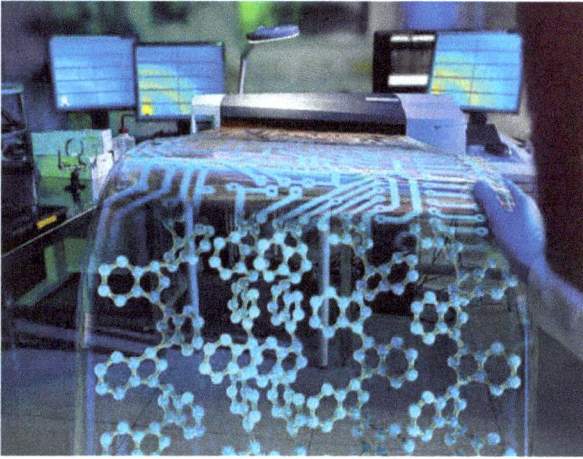

Figure 7.4: Polymers in electronics (image credit: https://als.lbl.gov/custom-organic-electronics-out-of-the-printer/).

repeated dosing. For example, biodegradable polymers such as polylactic-co-glycolic acid (PLGA) and polycaprolactone (PCL) are ideal for encapsulating drugs and gradually releasing them in the body, improving therapeutic outcomes. On the other hand, biodegradable polymers are used for temporary implants or medical devices that degrade in the body after performing their function. These polymers reduce the need for surgical removal after use and are ideal for applications like drug delivery, sutures, and bone fixation.

Figure 7.5: Polymers in medical devices (image credit: https://medicaldialogues.in/news/industry/medical-devices/medical-device-industry-needs-flexibility-in-labor-laws-tax-incentives-experts-67136).

7.2.5 Packaging

Polymers have transformed the packaging industry (Figure 7.6) by providing lightweight, cost-effective, and versatile materials for a variety of products. For example, polymers like polyethylene (PE), polypropylene (PP), and polyethylene terephthalate (PET) are commonly used for food and beverage packaging. These polymers are utilized in bottles, containers, films, and wraps. They offer excellent protection against moisture, air, and contaminants, thereby extending the shelf life of products. Polymers are also easily moldable, which makes them ideal for producing various shapes and sizes of containers. Polymer materials such as PVC, PET, and polycarbonate are commonly used in medical packaging. These polymers provide the required protection, sterilization compatibility, and mechanical strength necessary for medical applications. They can be molded into precise shapes to accommodate sensitive equipment and medications. The push for more sustainable packaging has also led to the development of biodegradable polymers and recycling technologies. Recycled PET (rPET) is increasingly used in food and beverage containers, and new biodegradable plastics, like PLA, are helping reduce the environmental impact.

Figure 7.6: Polymers in the packaging industry (image credit: http://www.packcon.org/index.php/en/articles/113-2017new/200-the-importance-of-plastic-polymers-in-packaging).

7.2.6 Textiles and Fibers

Polymers play a significant role in the textile industry (Figure 7.7), providing a wide range of fibers for clothing, industrial applications, and non-woven materials. Synthetic fibers such as nylon, polyester, acrylic, and polypropylene are widely used in clothing, upholstery, ropes, and industrial fabrics. Polymers are also utilized in the production of non-woven fabrics, which are extensively employed in hygiene products (diapers, sanitary pads), medical supplies (bandages, surgical gowns), and industrial applications (filters, insulation).

Figure 7.7: Polymers in the textile industry (image credit: https://www.iranpetroleum.co/the-use-of-polymer-materials-in-the-textile-industry/).

7.3 Essential Keywords

Esthetic design The focus on the visual appeal of the polymer product. This includes the texture, color, and overall appearance of the product.

Conductive polymers Organic materials capable of conducting electricity. They are used in electronic devices, batteries, and sensors due to their lightweight and flexible nature.

Cost optimization Is the cost-effectiveness of product design that designers need to follow in order to balance material costs, manufacturing costs, and functionality to produce a cost-efficient polymer product that meets both performance and budgetary requirements.

Design for manufacturability The designing of polymer products in a way that minimizes production costs, complexity, and manufacturing time. It involves considering factors such as material selection, process capabilities, and assembly requirements early in the design phase.

Durability and longevity Understanding the material's fatigue resistance, aging characteristics, and its response to environmental factors such as UV radiation, temperature fluctuations, and moisture.

Electronic polymers Materials used in electronics, such as OLED displays, solar cells, and organic semiconductors. These materials enable flexible, lightweight, and cost-effective electronic devices, which are key to the development of wearable electronics and flexible displays.

Energy storage polymers Materials that are used in energy storage technologies, such as batteries and supercapacitors.

Functionality and ergonomics The incorporating ergonomics into polymer product design to ensure that the product is comfortable and user-friendly, especially in applications like consumer electronics, medical devices, and tools.

Material selection Is the process of choosing the right polymer material to ensure optimal product performance, durability, and cost-effectiveness.

Medical devices polymers Materials are extensively utilized in the medical field for implants, prosthetics, drug delivery systems, and surgical instruments.

Polymeric textiles Synthetic polymers like polyester, nylon, and acrylic that are used in the textile industry for their durability, flexibility, and resistance to wear.

Polymer processing techniques Various polymer processing techniques, such as injection molding, extrusion, blow molding, and thermoforming.

Polymer recycling The converting of polymer materials back into usable products. Methods include mechanical recycling, chemical recycling, and biological recycling, with a growing focus on making recycling more efficient and eco-friendlier.

Smart polymers Materials that change their properties in response to external stimuli, such as temperature, pH, or light.

Structural integrity Designing a polymer product with the right structural integrity. This involves ensuring that the product can withstand the applied loads and environmental conditions without failure.

Weight reduction An advantage that can be achieved through material choice, part geometry, and optimizing the product design to use less material without compromising strength.

7.4 Questions and Answers

Questions	Answers
1. What is the role of material selection in polymer product design?	Material selection is one of the most critical aspects of polymer product design. It determines the product's mechanical properties, durability, thermal resistance, chemical resistance, and esthetic qualities. The right polymer must be chosen based on the product's intended function, environmental exposure, and production methods to ensure optimal performance.
2. How does polymer shrinkage affect product design?	Polymers tend to shrink as they cool after being molded, which can affect the final dimensions of the product. Designers must account for this shrinkage by adjusting the mold design, material choice, and processing conditions. Failure to compensate for shrinkage can result in dimensional inaccuracies and product defects.
3. What is design for manufacturability (DFM), and why is it important?	Design for manufacturability is the practice of creating polymer products with a focus on simplifying the manufacturing process. It involves selecting materials and design features that are easy to produce, thereby minimizing complexity, reducing cycle time, and lowering production costs. DFM ensures that the product design is not only functional and high-performing but also efficient, cost-effective, and compatible with existing manufacturing technologies.
4. How do tolerances impact polymer product design?	Tolerances are related to the allowable variation in a product's dimensions. Tight tolerances require high precision in the design and manufacturing process. It's crucial to determine the appropriate tolerance level based on the part's function and the capability of the manufacturing process to ensure proper fit and functionality.
5. What are the key considerations for designing lightweight polymer products?	Lightweight design is often achieved by selecting materials with low density, optimizing the geometry of the product to minimize material use, and integrating hollow sections or ribs for added strength without increasing weight. Lightweight polymer products are desirable in industries such as automotive and aerospace, where reducing weight improves efficiency and performance.

(continued)

Questions	Answers
6. Why is structural integrity important in polymer product design?	Structural integrity ensures that the polymer product can withstand the loads, stresses, and environmental conditions it will face in use. A product with poor structural integrity is likely to fail prematurely, so designers need to evaluate factors such as strength, stiffness, and fatigue resistance to ensure the product meets performance requirements without compromising safety.
7. How does sustainability influence polymer product design?	Sustainability in polymer product design involves using materials and manufacturing processes that minimize environmental impact. This can include selecting biodegradable or recyclable polymers, designing for product end-of-life recycling, and reducing waste and energy consumption during manufacturing. Sustainable design helps reduce the carbon footprint and supports a circular economy.
8. What is the significance of functional and ergonomic design in polymer products?	Functional and ergonomic design focuses on creating polymer products that are not only practical but also comfortable to use. By ensuring a product is easy to handle, intuitive, and free from strain or discomfort, user satisfaction and safety are significantly improved.
9. How are polymers used in the automotive industry?	In the automotive industry, polymers are extensively used to reduce vehicle weight, enhance fuel efficiency, and improve safety. Polymer-based materials are employed in the production of components such as bumpers, dashboards, interior trims, and engine parts.
10. How are polymers used in medical devices?	Polymers are used in medical devices due to their biocompatibility, flexibility, and ease of processing. They are utilized in products such as surgical instruments, implants, prosthetics, drug delivery systems, and catheters. Materials like silicone, polyurethane, and polyethylene are commonly employed for these applications.
11. How do polymers benefit the packaging industry?	The packaging business relies heavily on polymers due to their cost-effectiveness, adaptability, and superior barrier qualities against light, oxygen, and moisture. Polyethylene (PE), polypropylene (PP), and polyethylene terephthalate (PET) are materials that are often used in consumer items, medical supplies, and food packaging.

(continued)

Questions	Answers
12. How are polymers used in the textile industry?	In the textile industry, synthetic polymers like polyester, nylon, and acrylic are used to create fabrics for clothing, industrial textiles, and non-woven materials. Polymers provide durability, flexibility, and resistance to wear and are utilized in applications ranging from sportswear to automotive upholstery and medical textiles.
13. What are smart polymers, and where are they used?	Smart polymers are substances that can alter their structure, color, or chemical makeup in response to external stimuli such as light, pH, or temperature. These polymers provide cutting-edge solutions for consumer, medicinal, and environmental applications, including drug delivery systems, biosensors, self-healing materials, and smart fabrics.
14. What is the role of conductive polymers in electronics?	Conductive polymers are used in electronic applications due to their ability to conduct electricity, similar to metals, but with the added benefits of flexibility and lightweight properties. They are utilized in sensors, batteries, organic light-emitting diodes (OLEDs), and solar cells, enabling the development of flexible, lightweight, and cost-effective electronic devices.
15. How do mold design and polymer processing techniques affect product quality?	Techniques for polymer processing and mold design are essential to the end product's quality. A well-made mold guarantees uniform cooling, steady material flow, and minimal flaws. For the polymer product to have the appropriate shape, texture, and performance attributes, polymer processing methods, including blow molding, extrusion, and injection molding, must be optimized.

Chapter 8
Polymer Recycling, Upcycling, and Downcycling

Recycling is the act of breaking waste items into their basic component elements and then repurposing them to create new goods. Although it aids in waste reduction and resource conservation, material quality is frequently lost, particularly in conventional mechanical recycling. The practice of turning waste materials or discarded goods into new materials or products with improved quality, usefulness, or environmental value is known as upcycling. In contrast to traditional recycling, upcycling improves the material's qualities or repurposes it in a way that makes it more valuable rather than degrading it. Downcycling is a type of recycling where waste materials are converted into new products of lower quality and functionality compared to the original material. This happens when the recycling process degrades the material's properties, limiting its potential for future use. While recycling is an essential waste management strategy, it often results in material degradation over time. Upcycling, however, is the most sustainable approach when feasible, as it enhances the material without compromising its quality. Downcycling extends the life of waste but does not fully prevent it from being discarded. A brief comparison of upcycling, recycling, and downcycling is provided in Table 8.1.

Table 8.1: A brief comparison of recycling, upcycling, and downcycling.

Feature	Recycling	Upcycling	Downcycling
Material quality	Slightly reduced	Improved or maintained	Degraded
Energy requirement	Moderate to high	Lower	Moderate
Environmental impact	Reduces waste while significantly reducing energy consumption	Reduces waste with minimal processing	Delays waste but does not fully prevent disposal
Examples	PET bottles into new PET bottles with reduced quality	PET bottles to high-quality polyester fabric	PET bottles to plastic lumber

8.1 Polymer Recycling

Reducing plastic waste, preserving resources, and lessening the environmental impact of polymer products are all made possible through polymer recycling. Recycling reduces landfill waste and encourages the reuse of plastic products, both of which are important components of the circular economy. Technologies for recycling polymers are varied and constantly evolving. Chemical recycling presents a viable alternative

https://doi.org/10.1515/9783111585734-008

for more complex materials and mixed plastic waste, even though mechanical recycling remains the most popular technique due to its ease of use and cost-effectiveness. Biological recycling, though still in the research phase, holds great potential for environmentally friendly plastic waste degradation. Each recycling technology has its advantages and challenges, and a combination of these methods will likely be required to effectively address the global plastic waste problem and move toward a more sustainable circular economy. Another important aspect of recycling is biodegradable and biopolymer recycling. While there are challenges in terms of recycling efficiency, infrastructure, and contamination, advancements in sorting, composting, and chemical recycling technologies are helping to improve the recycling of these materials. Recycling biodegradable plastics presents unique challenges due to their biological degradation processes. Similar to traditional plastic recycling, biodegradable plastics like PLA or PHA can be mechanically recycled through shredding, washing, and reprocessing. However, due to their relatively low melting points and mechanical properties, they often require specialized equipment. In some cases, biodegradable polymers can be chemically recycled back into their monomers through hydrolysis, glycolysis, or other processes. This enables the restoration of the polymer to its virgin material form, which can then be repolymerized into new products. Unlike traditional plastic recycling, composting is an ideal recycling method for biodegradable polymers like PLA and PHA. In an industrial composting facility, microorganisms break down these polymers into natural elements, such as water, carbon dioxide, and biomass. Some biodegradable polymers can even be converted into energy through thermal processes like pyrolysis or gasification. This involves heating the plastics in the absence of oxygen to produce syngas or other energy-rich compounds. There are several challenges to overcome for better recycling of biodegradable and biopolymers. Some of these challenges are listed in Table 8.2.

Table 8.2: Biopolymer recycling challenges.

Challenge	Details
Contamination	The effective separation and recycling of biopolymers is challenging, as they are frequently combined with other plastics in waste streams. The effectiveness of recycling can also be impacted by contamination from food, inks, adhesives, and other substances.
Infrastructure limitations	Fewer systems are in place for the recycling of biodegradable and biopolymer-based products, and the majority of the current recycling infrastructure is designed to handle conventional plastics derived from petroleum. There are currently few facilities specifically designed to sort, process, and recycle biopolymers.
Degradation in natural environments	While biodegradable polymers are designed to degrade, they often require specific conditions for effective breakdown. If biodegradable polymers are not disposed of in the proper environment, they may not decompose as intended, leading to the accumulation of plastic waste.

Table 8.2 (continued)

Challenge	Details
Cost and market development	In general, the cost of producing biopolymers is higher than that of conventional plastics made from petroleum. Their widespread use and recyclability are hampered by the price of raw ingredients, manufacturing processes, and limited economies of scale.
Recycling efficiency	Some biopolymers degrade in such a way that their properties are compromised, making them unsuitable for direct recycling back into high-quality products. For example, some biodegradable plastics may lose strength or mechanical properties after being recycled multiple times.

To understand the effectiveness and constraints of plastic recycling, it is essential to explore the conventional methods used, including mechanical, chemical, and biological recovery processes. Below is a detailed discussion of these technologies.

8.1.1 Mechanical Recycling

Mechanical recycling refers to the process of physically breaking down used plastics into smaller pieces or pellets, which can then be remolded or reprocessed into new products. This is the most common and widely used form of polymer recycling. The process of mechanical recycling involves six steps, which are briefly presented in Figure 8.1.

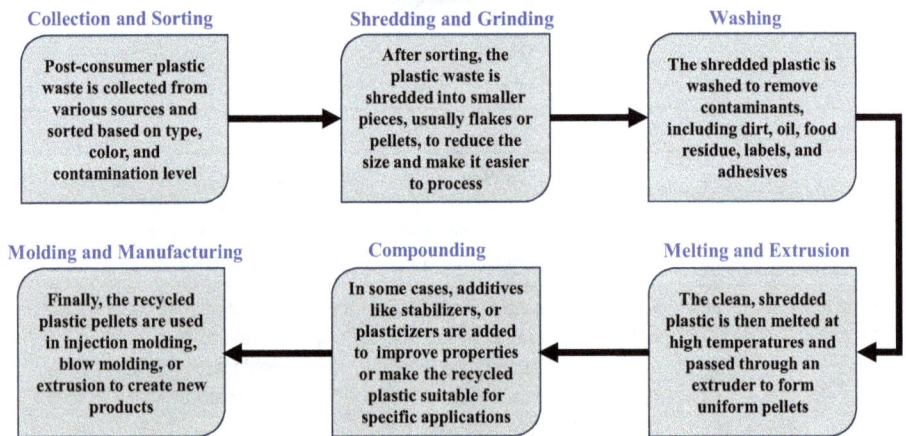

Collection and Sorting
Post-consumer plastic waste is collected from various sources and sorted based on type, color, and contamination level

Shredding and Grinding
After sorting, the plastic waste is shredded into smaller pieces, usually flakes or pellets, to reduce the size and make it easier to process

Washing
The shredded plastic is washed to remove contaminants, including dirt, oil, food residue, labels, and adhesives

Molding and Manufacturing
Finally, the recycled plastic pellets are used in injection molding, blow molding, or extrusion to create new products

Compounding
In some cases, additives like stabilizers, or plasticizers are added to improve properties or make the recycled plastic suitable for specific applications

Melting and Extrusion
The clean, shredded plastic is then melted at high temperatures and passed through an extruder to form uniform pellets

Figure 8.1: Schematic diagram of the mechanical recycling process.

8.1.2 Chemical Recycling

The technique of using chemical processes to degrade polymers into their component chemicals is known as chemical recycling. Compared to mechanical recycling, chemical recycling can handle mixed polymers and produce superior products that are often identical to the original plastic. Chemical recycling techniques are gaining popularity because they enable the recycling of polymers that are difficult to process mechanically. Chemical recycling techniques include solvolysis, depolymerization, gasification, and pyrolysis. These categories are explained in further detail in Figure 8.2.

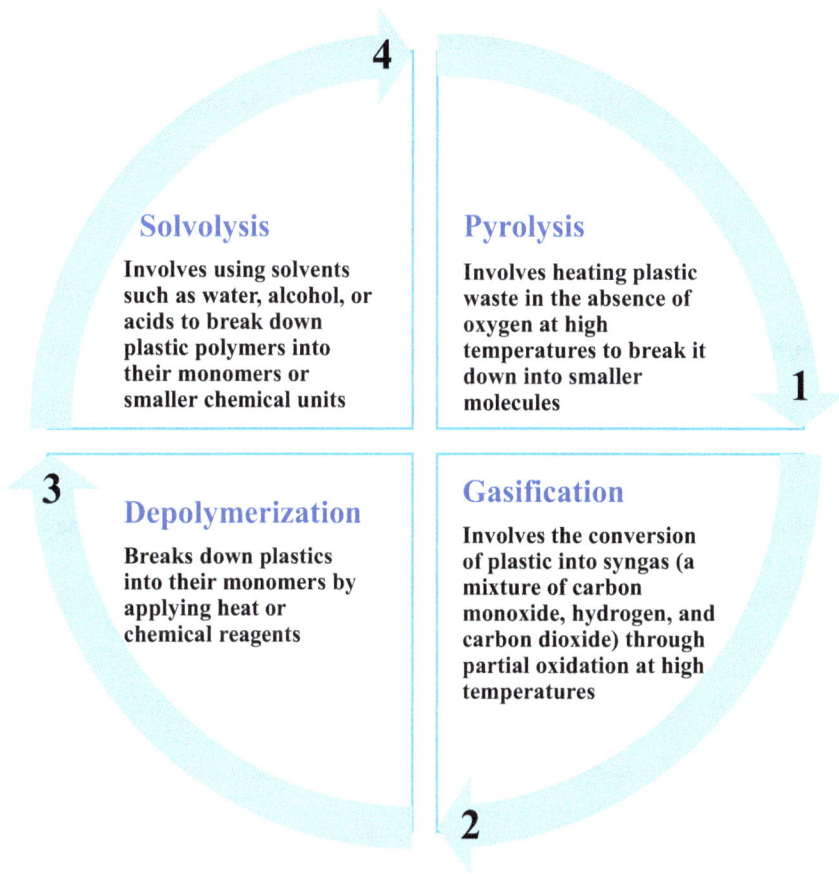

4

Solvolysis

Involves using solvents
such as water, alcohol, or
acids to break down
plastic polymers into
their monomers or
smaller chemical units

Pyrolysis

Involves heating plastic
waste in the absence of
oxygen at high
temperatures to break it
down into smaller
molecules

1

3

Depolymerization

Breaks down plastics
into their monomers by
applying heat or
chemical reagents

Gasification

Involves the conversion
of plastic into syngas (a
mixture of carbon
monoxide, hydrogen, and
carbon dioxide) through
partial oxidation at high
temperatures

2

Figure 8.2: Schematic diagram of the chemical recycling types.

8.1.3 Biological Recycling

Biodegradation, also known as biological recycling, is the process of decomposing plastics through biological agents such as fungi, bacteria, and enzymes. This approach is primarily applicable to biodegradable polymers and remains in the early stages of development. Biodegradable plastics, including polylactic acid (PLA), polyhydroxyal-kanoates (PHA), and starch-based polymers, can be broken down by microorganisms into biomass, carbon dioxide, and water. In biological recycling, microorganisms naturally degrade polymer chains by producing enzymes that cleave the bonds between monomers. Among the limitations of biological recycling are:

- Slow Process: Biological degradation can be slow and may require specific environmental conditions (e.g., temperature, humidity, and oxygen levels) for efficient breakdown.
- Limited to Biodegradable Plastics: Most of the plastics used today, such as PE, PP, and PET, are not biodegradable and cannot be processed through biological recycling.
- Limited Commercial Scale: The technology is still at an experimental stage and is not yet scalable for large-scale plastic waste management.

8.2 Polymer Upcycling

Polymer upcycling involves the selective breakdown of polymers into chemicals, fuels, or molecular intermediates, followed by their reassembly into high-value products under mild conditions. Essentially, upcycling is a form of creative reuse where discarded materials, waste, or otherwise low-value products are transformed into items of greater quality or utility.

Unlike traditional recycling, upcycling often takes place at the manufacturing stage rather than in post-consumer settings, enabling significant material and cost savings at an industrial scale. This advanced approach to plastic waste management enhances or modifies polymers by incorporating additives, reinforcements, or blending them with other materials to create more functional, durable, and valuable products. Polymer upcycling aligns with sustainability and circular economy principles, aiming to reduce the environmental impact of plastic waste while improving resource efficiency.

8.2.1 Types of Polymer Upcycling

Polymer upcycling can be classified into mechanical, chemical, and biological upcycling, each offering unique advantages (Table 8.3).

Table 8.3: Mechanical, chemical, and biological upcycling.

Mechanical	Chemical	Biological
The key strategies in mechanical upcycling are: – Mixing waste polymers with high-performance polymers to improve mechanical properties. – Incorporating nanoparticles (e.g., graphene, carbon nanotubes, nanocellulose) to enhance strength, conductivity, and thermal stability. – Functionalizing polymer surfaces to increase adhesion, hydrophobicity, or compatibility with other materials.	Chemical upcycling involves breaking down polymer chains and reassembling them into high-value products. This includes: – Using catalysts to selectively break down polymers into monomers or oligomers. – Reacting polymers with solvents to produce raw materials for new applications. – Modifying polymer chains to introduce functional groups enhances material properties.	Biological upcycling utilizes enzymes and microorganisms to degrade and reassemble polymers into valuable bioproducts. This includes: – Breaking down polymers into monomers using specialized enzymes. – Transforming degradation products into biofuels, biodegradable plastics, or specialty chemicals.

8.2.2 Applications of Polymer Upcycling

Upcycled polymers have diverse applications across multiple industries. This includes:

– High-performance materials, such as carbon fiber-reinforced plastics (CFRPs) made from upcycled polymers, improve strength-to-weight ratios for aerospace and automotive applications, as well as self-healing polymers designed for biomedical and electronic applications.
– Sustainable textiles and fibers, such as upcycled PET fibers used in clothing, carpets, and upholstery, and biodegradable polymer coatings that enhance fabric durability while reducing microplastic pollution.
– Upcycled conductive polymers that enable flexible electronics, batteries, smart coatings, and photoactive upcycled polymers contribute to solar cell efficiency.
– Upcycled polymer adsorbents that remove heavy metals, oil spills, and organic contaminants from wastewater, along with upcycled polymer membranes used in filtration technologies for clean water access.

8.2.3 Recent Advances and Innovations

Several novel technologies have accelerated the development of polymer upcycling. Examples include:
–   Artificial Intelligence (AI) and Machine Learning (ML): Optimizing upcycling processes by predicting reaction conditions and material properties.
–   3D Printing with Upcycled Polymers: Repurposing waste polymers for additive manufacturing.
–   Plasma-Based Upcycling: Using plasma treatment to introduce functional groups into waste polymers, thereby enhancing their reusability.
–   Biorefinery Integration: Converting waste polymers into biofuels and biochemical precursors.

8.3 Polymer Downcycling

Polymer downcycling is the process of converting plastic waste into lower-value materials with reduced mechanical and chemical properties compared to the original polymer. Unlike upcycling, which enhances the value of waste polymers, downcycling degrades material quality, limiting the number of recycling cycles before the polymer becomes unusable. Although downcycling helps divert plastic waste from landfills and incineration, it is not a long-term sustainable solution due to the progressive loss of material properties. In many cases, it temporarily extends the life of plastics that would otherwise be discarded. A debated topic in the waste and recycling industry, some argue that all plastic recycling is essentially downcycling, as post-consumer plastics often lack the structural integrity of the original materials. In summary, while polymer downcycling contributes to waste reduction, it results in material degradation and restricts recycling potential. Unlike upcycling, which increases value, downcycling produces lower-quality materials with a limited number of reuse cycles.

8.3.1 Types of Polymer Downcycling

Downcycling processes typically fall under mechanical and chemical methods, with mechanical methods being the most common (Table 8.4). Examples of mechanically and chemically downcycled polymers are also shown in Figure 8.3.

8.3.2 Applications of Downcycled Polymers

Downcycled polymers are often used in non-durable, low-performance applications where material degradation is less critical. Some common applications include:

Table 8.4: Comparison of mechanical and chemical downcycling processes.

Mechanical	Chemical
Mechanical downcycling refers to the physical reprocessing of plastic waste while preserving the polymer's chemical structure. However, this process typically results in reduced material properties with each recycling cycle. The main methods of mechanical downcycling include: – Grinding and Pelletizing: Waste plastics are shredded, melted, and reformed into pellets for reuse, although the resulting material has diminished mechanical strength. – Filler Addition: Inorganic fillers, such as calcium carbonate or talc, are incorporated to compensate for strength loss; however, this further limits recyclability. – Blending with Other Polymers: Recycled polymers are mixed with other materials to restore some properties; however, degradation continues with successive cycles.	Chemical downcycling involves partial depolymerization of plastics into lower-value oligomers or secondary chemicals, which are then repurposed. However, the recovered materials often have lower purity and usability. Processes include: – Pyrolysis: Plastics are thermally decomposed into waxes, oils, and gaseous hydrocarbons; however, these products are generally lower in value compared to the original polymer. – Hydrolysis and Alcoholysis: PET and polyesters can be broken down into lower-value oligomers, which are then used as binders, adhesives, or low-grade resins. – Oxidative Degradation: Used to convert plastics into simple hydrocarbons; however, the resulting compounds are typically used in lower-value applications, such as fuel additives.

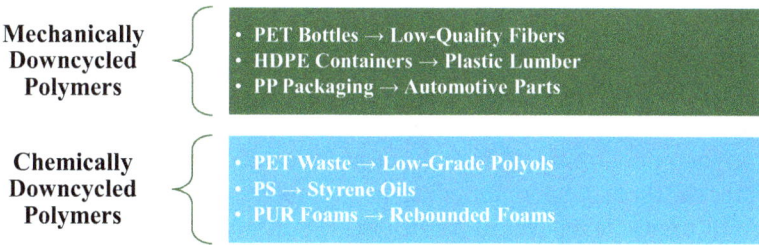

Mechanically Downcycled Polymers
- PET Bottles → Low-Quality Fibers
- HDPE Containers → Plastic Lumber
- PP Packaging → Automotive Parts

Chemically Downcycled Polymers
- PET Waste → Low-Grade Polyols
- PS → Styrene Oils
- PUR Foams → Rebounded Foams

Figure 8.3: Examples of mechanically and chemically downcycled polymers.

– **Construction and Infrastructure**
  – Plastic Lumber: Made from downcycled HDPE and PP, it is used for decking and fencing.
  – Composite Panels: Low-grade polymer composites are used in insulation, roofing, and siding.
– **Textile and Fiber Products**
  – Non-woven Fabrics: Downcycled PET fibers are used in disposable items such as wipes and geotextiles.
  – Insulation Materials: Low-value PET fibers are repurposed for sound and thermal insulation.

- **Automotive and Industrial Uses**
  - Recycled Plastic Components: Used in low-stress automotive parts, such as interior trims and fender liners.
  - Pallets and Shipping Containers: Made from downgraded polypropylene and polyethylene.
- **Packaging and Consumer Goods**
  - Trash Bags and Plastic Films: Made from downgraded polyethylene.
  - Cheap Plastic Furniture: Often brittle and prone to cracking over time.

8.3.3 Limitations and Environmental Concerns

Despite its role in waste management, polymer downcycling poses several challenges:
- Material Property Decline: Each downcycling cycle shortens polymer chains, reducing their usability.
- Limited Market Demand: Downcycled products often have low commercial value.
- Shorter Lifespan: Many downcycled products degrade quickly and eventually end up in landfills.
- Microplastic Generation: Reprocessing weakens polymer structures, leading to fragmentation and microplastic pollution.

8.4 Plastic Recycling Symbol Codes and Numbers

Plastic recycling symbol codes (Figure 8.4) or SPI (Society of the Plastics Industry) codes (Figure 8.5) serve to identify the material composition of plastic items, aiding in the recycling process. These symbols, including the chasing arrows logo or resin identification code (RIC), indicate the type of plastic used but do not necessarily mean the material is recyclable. Instead, they provide information about the polymer composition, helping recyclers determine the appropriate processing method.

8.5 Essential Keywords

Biological recycling A type that uses microorganisms or enzymes to break down polymers, typically biodegradable plastics, into simpler, non-toxic compounds. This process is often slower but can be very effective for biodegradable polymers like PHA.

Chemical recycling A process that breaks down polymers into their original monomers or other valuable chemicals. This process allows for the recycling of thermosets and contaminated polymers, enabling them to be used for the production of new polymers or chemicals.

PET or PETE (Polyethylene terephthalate). It is used to make bottles for soda, water and other drinks. It's also used to make cooking oil containers, plastic peanut butter jars and containers for other popular food items. PET/PETE products CAN be recycled.

HDPE (High density polyethylene). It is used to make milk jugs, shampoo bottles, cleaning product containers and detergent bottles. HDPE products CAN be recycled.

PVC (Polyvinyl chloride). It is used for a huge array of household products. Plastic tubing, kids' toys, plastic trays and furniture are often made from PVC. PVC products CANNOT be recycled.

LDPE (Low density polyethylene). It is used to make grocery bags and the bags that hold newspapers, sliced bread loaves and fresh produce, among other things. LDPE products CAN SOMETIMES be recycled.

PP (Polypropylene). It is used to make the food containers used for products like yogurt, sour cream and margarine. It's also made into straws, rope, carpet and bottle caps. PP products CAN SOMETIMES be recycled.

PS (Polystyrene). It is commonly used to make disposable coffee cups, packing peanuts, coolers and to-go food containers. PS products CAN SOMETIMES be recycled.

Other. Any type of plastic that doesn't fit into one of the first six categories falls under this heading. Products stamped with a 7 are often made from multiple plastic types or out of other types of plastic that can't easily be recycled. Common products that use # 7 plastics are anything marked as 'bioplastic' or 'compostable' plastic, Baby bottles and sippy cups, water cooler jugs, and car parts.

Figure 8.4: Plastic recycling symbols, codes, and numbers.

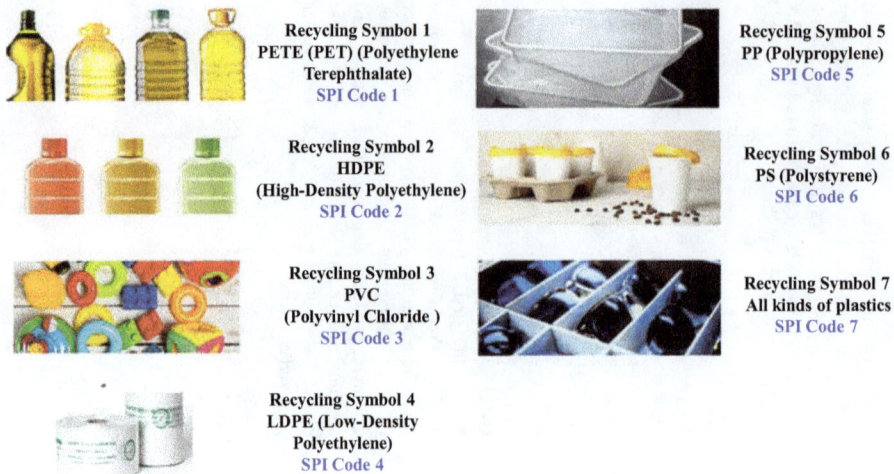

Recycling Symbol 1 PETE (PET) (Polyethylene Terephthalate) SPI Code 1	**Recycling Symbol 5** PP (Polypropylene) SPI Code 5
Recycling Symbol 2 HDPE (High-Density Polyethylene) SPI Code 2	**Recycling Symbol 6** PS (Polystyrene) SPI Code 6
Recycling Symbol 3 PVC (Polyvinyl Chloride) SPI Code 3	**Recycling Symbol 7** All kinds of plastics SPI Code 7
Recycling Symbol 4 LDPE (Low-Density Polyethylene) SPI Code 4	

Figure 8.5: Society of the plastics industry (SPI) codes with examples (image credit: https://www.biogone.com.au/news/recycling-codes-1-7/).

Chemical recycling of biopolymers Refers to the process of recycling biopolymers back to their original monomer form, which can then be used to produce new biopolymers or other chemicals. This process helps reduce the need for raw resources.

Closed-loop recycling Is the process of recycling polymers back to their original form or using polymers to create the same type of product again. This approach is ideal for materials like PET, where recycled bottles are reused to make new bottles.

Downcycling The process of recycling polymers into products of lower quality or value, often due to degradation in material properties after repeated cycles. For example, PET bottles may be recycled into lower-quality textiles or insulation materials.

Enzyme-based biopolymer recycling Is an emerging technique where enzymes are used to break down biopolymers into simpler monomers or oligomers, which can then be repurposed for new materials. This method is often used for biodegradable plastics like PHA.

Mechanical recycling A process that involves shredding, melting, and reforming polymers to reuse them in new products without altering their chemical structure.

Open-loop recycling is a process that involves recycling polymers into a different type of product. For example, recycled PET could be used to make clothing fabrics, reducing waste but not returning the material to its original form.

Pyrolysis A thermal degradation process in which polymers are heated without oxygen, breaking them down into smaller molecules that can be used as fuels or feedstock for new materials.

Solvent-based recycling A technique that dissolves polymers in a solvent to separate them from contaminants. Once purified, the polymer can be reprocessed or repurposed into new products.

Upcycling The transformation of waste materials into new products with enhanced value or quality.

8.6 Questions and Answers

Questions	Answers
1. What is polymer recycling?	Polymer recycling involves the collection and processing of polymer waste to create new products, reducing the need for original materials and minimizing environmental impact. The process can be mechanical, chemical, or biological, depending on the type of polymer and the desired product.
2. What are the main benefits of recycling?	Recycling reduces landfill waste, conserves natural resources, saves energy, and lowers pollution.
3. What is mechanical recycling of polymers?	Mechanical recycling involves physically breaking down used polymer products (such as plastics) into smaller pieces, melting them, and reforming them into new products.
4. What is chemical recycling, and how does it differ from mechanical recycling?	Chemical recycling breaks down polymers into their original monomers or other useful chemicals through chemical processes such as pyrolysis or hydrolysis. Unlike mechanical recycling, which simply melts and reshapes materials, chemical recycling enables the reuse of thermosets and more contaminated polymers by restoring their original chemical structure.
5. Can biodegradable polymers be recycled?	Yes, biodegradable polymers, such as PLA (polylactic acid) and PHA (polyhydroxyalkanoates), can be recycled, although the process is more complex than recycling traditional plastics. Biodegradable polymers may require specialized collection, sorting, and processing methods.

(continued)

Questions	Answers
6. What is the role of enzyme-based recycling in biopolymer recycling?	Enzyme-based recycling uses specific enzymes to break down biodegradable polymers into their constituent monomers or oligomers. This process enables the recycling of biopolymers like PHA and PLA, facilitating their reuse for new materials or chemicals and improving the efficiency of biopolymer recycling.
7. What are the advantages of using chemical recycling for biopolymers?	By breaking down into their original monomers, biopolymers can be chemically recycled and used to make new biopolymers or other compounds. This procedure can help develop a more sustainable, circular economy for biopolymer goods, reduce waste, and lessen the demand for fresh raw materials.
8. What is the difference between upcycling and downcycling in polymer recycling?	Upcycling refers to the process of turning waste polymers into products of higher quality or value, such as creating durable composites from recycled plastics. Downcycling, on the other hand, involves turning waste polymers into products of lower quality or value, such as turning PET bottles into insulation materials or lower-grade plastics.
9. What are the environmental benefits of biodegradable polymers?	Biodegradable polymers decompose into non-toxic compounds, in contrast to conventional plastics, which might remain in landfills for hundreds of years. This, in turn, reduces long-term plastic pollution.
10. Can biopolymer recycling be integrated into a circular economy?	Yes, biopolymer recycling plays a key role in the circular economy. Biopolymers like PLA and PHA can be recycled into new products, reducing reliance on fossil resources. They offer sustainable alternatives to traditional plastics, promoting reuse and reducing plastic waste in the environment.
11. What are some challenges associated with recycling biodegradable polymers?	Challenges with recycling biodegradable polymers include the need for specialized infrastructure to collect, sort, and process these materials. Additionally, biodegradable polymers may require extra care during recycling due to possible contamination with non-biodegradable plastics.

(continued)

Questions	Answers
12. Why is downcycling less sustainable than upcycling?	Downcycling reduces the quality of materials over time, leading to eventual waste, whereas upcycling extends the usability of materials without degrading them.
13. What role does consumer behavior play in recycling success?	Proper waste segregation and responsible disposal by consumers improve recycling efficiency and reduce contamination.
14. Why is upcycling considered environmentally friendly?	Upcycling minimizes the need for new raw materials and reduces energy consumption compared to traditional recycling.
15. Can glass be upcycled?	Yes, glass bottles can be upcycled into other materials, such as lamps, decorative items, or reusable containers.

Chapter 9
Polymers Environmental and Societal Impact

Plastics, a prominent class of polymers, have significantly improved everyday life through their convenience, affordability, and diverse applications. Nevertheless, the environmental burden they impose, including pollution, dwindling natural resources, and rising greenhouse gas emissions, is increasingly troubling. Beyond environmental damage, plastic waste poses public health risks, exacerbates social inequalities, and contributes to long-term ecological degradation. Addressing these multifaceted challenges demands an integrated approach that blends technological innovation, eco-conscious design, more efficient recycling practices, and robust policy measures. Such efforts are essential to advancing a circular economy and mitigating the harmful impact of polymers on society and the environment.

Polymers, due to their widespread use in countless applications, have had profound effects on both the environment and society. While polymers, particularly plastics, have brought numerous benefits in terms of convenience, durability, and cost-effectiveness, their environmental impact has become a major concern. In this chapter, the environmental impact of polymers and their societal implications, both positive and negative, are briefly explored.

9.1 Environmental Impact of Polymers

The environmental impact of polymers largely stems from their production, use, and disposal. The most commonly discussed polymers in this context are synthetic plastics, but biopolymers and biodegradable plastics also have their unique environmental profiles.

9.1.1 Plastic Pollution and Waste

- Landfill Accumulation: Synthetic polymers, especially those used in packaging, single-use items, and consumer products, often end up in landfills after disposal. Due to the enormous volume of plastic waste generated each day and the non-biodegradable nature of most plastics, this waste can remain in the environment for centuries.
    - Non-biodegradable Nature: Conventional plastics such as polyethylene (PE), polypropylene (PP), and polyethylene terephthalate (PET) are resistant to natural degradation. As a result, they accumulate over time, contributing significantly to landfill congestion and long-term environmental pollution.

https://doi.org/10.1515/9783111585734-009

- Microplastics Formation: As larger plastic items slowly degrade, they fragment into microscopic particles known as microplastics. These particles are now widespread in marine and terrestrial environments, including oceans, rivers, soil, and even the atmosphere, posing serious threats to ecosystems and wildlife.
- Marine Pollution: It involves the contamination of oceans, seas, and coastal regions by harmful substances such as plastics, chemicals, oil, and industrial waste. This form of pollution poses a significant threat to marine ecosystems, leading to habitat degradation and loss of biodiversity. Plastic waste is particularly problematic, causing injury or death to marine animals through ingestion and entanglement. Chemical contaminants, including heavy metals and pesticides, accumulate in aquatic food chains, endangering marine life and posing risks to human health. Oil spills result in long-lasting ecological damage, while untreated sewage and agricultural runoff further degrade water quality. Addressing marine pollution necessitates coordinated global action, encompassing stricter environmental regulations, improved waste management systems, and increased public awareness and education.
  - Ingestion by Marine Life: A significant portion of plastic waste ends up in the oceans, where it is ingested by marine life, from small plankton to large whales. This leads to physical harm, toxicity, and even death for marine organisms. It also disrupts the food chain, as plastics can carry toxic chemicals or enter the tissues of marine animals.
  - Plastic Islands: Large accumulations of plastic debris, such as the Great Pacific Garbage Patch, are examples of how polymers can form concentrated, floating islands of waste, creating ecological dead zones and endangering biodiversity.
- Soil Contamination: Plastic particles can infiltrate soil systems, negatively impacting soil health, structure, and fertility. As plastics degrade, they may release toxic additives and chemicals into the soil, which can harm plants, microorganisms, and other organisms that depend on the land for survival.
- Air Pollution during Incineration: Incinerating plastic waste to reduce its volume can lead to the emission of hazardous pollutants such as carbon dioxide (CO_2), carbon monoxide (CO), dioxins, and furans. These substances contribute to air pollution and pose serious health risks to both humans and animals, as well as causing environmental degradation.

9.1.2 Resource Depletion

The production of synthetic polymers relies heavily on fossil fuels, particularly petroleum. This contributes to the depletion of non-renewable resources and increases the environmental footprint of polymer manufacturing.

- Fossil Fuel Consumption: Polymers like PE, PP, and PVC are derived from petrochemicals, which require the extraction, refining, and processing of fossil fuels. This process contributes significantly to global greenhouse gas emissions, resource depletion, and environmental degradation.
- Energy-Intensive Production: Polymer production can be energy-intensive, particularly during processes such as polymerization, extrusion, and molding. The energy consumed during polymer manufacturing contributes to carbon emissions and environmental impact.

9.1.3 Greenhouse Gas Emissions and Climate Change

- Carbon Footprint: The production of polymers, from raw material extraction through polymerization, manufacturing, and distribution, generates significant greenhouse gas emissions. These emissions play a major role in global warming and climate change, positioning polymers as a key contributor to the environmental footprint of industrial activity.
- Methane Emissions from Landfill Disposal: In landfills, certain plastics can break down under anaerobic conditions, releasing methane, a greenhouse gas considerably more potent than carbon dioxide. The large volume of plastic waste ending up in landfills intensifies this problem, further accelerating climate-related impacts.

9.2 Societal Impact of Polymers

Polymers have had an undeniable influence on modern society, both positively and negatively. They have transformed industries, improved the quality of life, and provided new solutions to societal challenges. However, the overreliance on polymers, particularly plastics, has led to various environmental and societal issues.

9.2.1 Positive Societal Impacts

Positive societal impacts include improved quality of life, economic growth, and environmental sustainability. Advancements in education, healthcare, and technology enhance well-being and create opportunities for individuals and communities. Social initiatives, such as recycling programs and clean energy adoption, promote sustainability and reduce environmental harm. Inclusivity, ethical practices, and innovation contribute to a more equitable and progressive society. These impacts are particularly evident in several key areas.

- Convenience and Affordability: Polymers have revolutionized consumer products by providing lightweight, durable, and cost-effective materials for a wide range of applications. Items such as packaging, electronics, and medical devices are often made from polymers, offering convenience and affordability that would otherwise be impossible with traditional materials like glass or metal. Polymers have dramatically reduced the weight of packaging materials, making goods cheaper to transport. Plastics are also used in food packaging, preserving freshness and extending shelf life, thus reducing food waste.
- Medical and Healthcare Innovations: Polymers have had a significant positive impact on healthcare, with applications ranging from disposable medical devices (e.g., syringes, gloves, and catheters) to advanced biomaterials used in implants, drug delivery systems, and wound dressings. Biodegradable polymers like PLA and PHA are increasingly used in medical applications, offering safe alternatives for temporary medical implants and drug delivery systems that naturally degrade in the body.
- Construction and Infrastructure: In construction, polymers are used in plumbing, insulation, roofing, and flooring materials, enhancing the durability, performance, and energy efficiency of buildings. Additionally, polymers contribute to the development of lightweight, strong composites utilized in building materials.
- Textiles and Clothing: Synthetic fibers like nylon, polyester, and spandex have transformed the fashion and textile industries. Polymers offer versatility in fabric design, durability, and moisture resistance, contributing to both everyday clothing and specialized performance fabrics used in sportswear, medical uniforms, and workwear.
- Automotive Industry: The automotive industry uses polymers extensively in manufacturing parts, ranging from lightweight plastic body components to advanced composites utilized in vehicle interiors and exteriors. Polymers help reduce vehicle weight, thereby improving fuel efficiency and lowering emissions.

9.2.2 Negative Societal Impacts

Negative societal impacts include environmental degradation, economic inequality, and social unrest. Pollution, resource depletion, and climate change threaten ecosystems and public health. Economic disparities can lead to poverty, unemployment, and limited access to education and healthcare. Technological advancements, while beneficial, may contribute to job displacement and privacy concerns. Addressing these challenges requires sustainable policies, ethical practices, and social responsibility. Here are some key areas where these impacts are evident.

- Plastic Waste Crisis: The overuse of plastic products, combined with inadequate waste management systems, has resulted in a global plastic waste crisis. Single-use plastics, such as bags, straws, and packaging, contribute significantly to litter, environmental degradation, and the clogging of drainage systems, leading to urban flooding.

– Health Risks and Toxicity: Certain polymers, especially those used in food packaging, medical devices, and consumer products – may contain hazardous additives such as phthalates, bisphenol A (BPA), and flame retardants. These substances can leach from the plastic into food, beverages, or the human body, posing risks such as endocrine disruption, developmental issues, and increased cancer potential. Additionally, the widespread presence of microplastics in the environment raises growing health concerns, as humans may ingest these particles through contaminated seafood, drinking water, or even air. While research into the long-term health impacts of microplastic exposure is still ongoing, early findings suggest potential risks to human health.

– E-Waste and Non-recyclable Polymers: Due in part to the widespread use of polymers in consumer electronics, the increasing amount of electronic waste, or "e-waste," poses a rising social and environmental challenge. Many of these polymer-based items end up in landfills and unofficial dumping sites because they are either non-recyclable or present major recycling issues. Hazardous materials, including heavy metals and toxic compounds, can leak into the environment as a result of improper e-waste disposal and processing, putting ecosystems and public health at risk.

– Social Inequity in Waste Management: The impact of plastic pollution disproportionately affects poorer communities and developing countries, where waste management infrastructure is often lacking. In these areas, plastic waste accumulates in the streets and waterways, posing severe health risks and environmental challenges. Furthermore, these communities often bear the brunt of plastic waste imports from developed nations.

9.3 Addressing the Challenges: Solutions and Strategies

To mitigate the environmental and societal impacts of polymers, several strategies and innovations are being explored:

– Improved Recycling Technologies: Developing more efficient recycling technologies for both conventional and biodegradable polymers will help reduce the environmental burden of plastic waste. Innovations in chemical recycling, such as depolymerization and pyrolysis, allow more plastics to be recycled into new products.

– Biodegradable and Compostable Alternatives: Promoting the use of biodegradable and compostable polymers in products such as food containers and packaging offers a practical approach to reducing plastic waste in landfills. However, the effectiveness of this strategy relies heavily on the presence of appropriate waste management systems – particularly industrial composting facilities capable of processing these materials under controlled conditions.

– Circular Economy and Sustainable Design: Advancing a circular economy, where products are intentionally designed for reuse, repair, and recyclability, is essential for minimizing the environmental footprint of polymers. By integrating sustainable design principles and encouraging industries to adopt eco-friendly manufacturing and end-of-life strategies, the overall lifecycle and utility of polymer-based products can be significantly extended.

– Consumer Education and Policy Reform: Raising public awareness about the environmental impacts of polymer waste and promoting responsible disposal and recycling practices are essential components of sustainable waste management. Meaningful progress can be achieved through strong governmental and corporate actions, including the implementation of single-use plastic bans, the promotion of eco-friendly packaging alternatives, and increased investment in efficient and accessible waste management infrastructure.

– Development of Biopolymers: The development and commercial scaling of biopolymers, such as PLA and PHA, derived from renewable resources, offers a more sustainable alternative to petroleum-based polymers. These materials are biodegradable, reducing the long-term environmental impacts associated with traditional plastics.

9.4 Essential Keywords

Biodegradability The capacity of a polymer to naturally break down into non-toxic byproducts when exposed to environmental elements, including light, heat, moisture, and microbial activity, is known as biodegradability.

Bioplastics Are polymers made from cellulose, sugarcane, and maize starch, among other renewable biological resources. They are intended to lessen reliance on fossil fuels and the carbon emissions that come with the conventional manufacturing of plastic.

Carbon footprint of polymers Refers to the total greenhouse gas emissions generated throughout the polymer's life cycle, from production to disposal. Many conventional polymers are petroleum-based, contributing to high carbon emissions, which impact climate change.

Circular economy A model of production and consumption designed to reduce waste and keep materials in use by promoting the recycling, reusing, and repurposing of products and materials. In the context of polymers, it focuses on developing systems where plastics are recycled into new products, thereby reducing the need for original resources.

Eco-friendly polymers Include biodegradable plastics, bio-based polymers, and those designed for easier recycling or composting. Examples include PLA (polylactic acid) and PHA (polyhydroxyalkanoates), which are derived from renewable resources and degrade more easily in the environment.

End-of-life solutions (EOL) Address how polymers are managed after completing their intended use. Common approaches include mechanical recycling, where plastics are physically reprocessed into new products; chemical recycling, which breaks polymers down into their monomers for reuse; energy recovery, involving the incineration of plastics to generate energy; and landfilling, the least desirable option due to its environmental impact. The overarching goal of EOL strategies is to minimize ecological harm while maximizing the recovery and reuse of valuable resources.

Environmental persistence Many conventional polymers are highly resistant to degradation, allowing them to persist in the environment for centuries. This, in turn, contributes to severe environmental challenges, especially in marine ecosystems, where accumulated plastic waste endangers wildlife, disrupts food chains, and damages habitats.

Extended producer responsibility (EPR) A policy framework that assigns manufacturers responsibility for the entire lifecycle of their products, including post-consumer waste management.

Life cycle assessment (LCA) A process of evaluating the environmental impact of a polymer from its production to its disposal.

Microplastics Are small plastic particles that result from the breakdown of larger plastic items or are intentionally manufactured for use in products such as cosmetics, cleaning agents, and textiles.

Plastic pollution The accumulation of plastic products and materials in the environment leads to environmental degradation, harm to wildlife, and ecosystem disruption.

Polymeric additives and toxins Polymeric additives, such as plasticizers, flame retardants, and stabilizers, are often used to enhance the properties of polymers but can pose environmental and health risks.

Resource depletion Continuous production of polymers from petroleum and natural gas contributes to resource depletion.

Sustainable packaging The use of biodegradable, recyclable, or reusable polymers for packaging aims to minimize plastic waste.

Waste-to-energy (WTE) Technologies convert non-recyclable polymer waste into energy, such as electricity or heat, through combustion, pyrolysis, or other thermal processes.

9.5 Questions and Answers

Questions	Answers
1. What is the environmental impact of polymers?	Polymers, particularly conventional plastics, pose serious environmental challenges due to their long-term persistence in natural ecosystems. A major consequence is plastic pollution, especially in marine environments, where accumulated debris endangers wildlife, disrupts ecosystems, and contributes to the degradation of biodiversity.
2. How do microplastics affect the environment?	Microplastics are small plastic particles that originate either from intentional production or from the degradation of larger plastic items. They contaminate the air, water, and soil, posing environmental hazards. Marine organisms can ingest these particles, leading to bioaccumulation and potentially allowing them to enter the food chain, thereby threatening the health of both humans and animals.
3. What are biodegradable polymers, and how do they benefit the environment?	Polymers that break down when exposed to environmental conditions. They reduce long-term pollution in landfills and oceans, as they decompose into non-toxic byproducts.
4. What is the concept of a circular economy in relation to polymers?	The concept aims to keep resources (polymers and plastics) in use for as long as possible by encouraging recycling, reuse, and repurposing of products.
5. How does plastic pollution impact marine life?	Plastic pollution poses a severe threat to marine life, with animals often ingesting plastic debris, which can lead to injury and/or death.
6. What is the role of life cycle assessment (LCA) in understanding the environmental impacts of polymers?	Life cycle assessment (LCA) evaluates the environmental impact of a polymer throughout its life cycle, from raw material extraction to disposal. It helps identify key areas for improvement in polymer production processes, energy consumption, waste management, and material sourcing.
7. Can bioplastics solve the plastic pollution problem?	Bioplastics, which are made from renewable resources and are biodegradable, still require specific conditions for degradation and may not break down effectively in certain environments, such as oceans or landfills.

(continued)

Questions	Answers
8. What are the societal implications of the increasing use of biodegradable plastics?	The use of biodegradable plastics can reduce plastic pollution. However, it also raises concerns about their proper disposal, the need for industrial composting infrastructure, and their potential impact on food production.
9. How does polymer recycling contribute to sustainability?	Polymer recycling reduces the need for raw materials, conserves energy, reduces waste, and lowers carbon emissions.
10. What are some challenges in reducing the environmental impact of polymers?	Challenges include improving recycling rates, designing more recyclable polymers, addressing the persistence of non-biodegradable plastics, and developing alternative materials that are both environmentally friendly and economically viable.
11. What is one way society can reduce polymer waste?	By using reusable products and supporting better recycling programs, we can contribute to a more sustainable future.
12. How does polymer production affect air quality?	It releases greenhouse gases and toxic chemicals during manufacturing.
13. Can biopolymers replace traditional plastics?	Yes, biopolymers offer an eco-friendly alternative with a lower environmental impact.
14. How can polymers be made more sustainable?	By using biodegradable materials and improving recycling methods, we can contribute to a more sustainable future.
15. How can individuals help reduce plastic pollution?	By reducing plastic use, recycling properly, and choosing biodegradable alternatives, we can make a positive impact on the environment.

Abbreviations

ABS	Acrylonitrile butadiene styrene
AEMs	Anion exchange membranes
AFM	Atomic Force Microscopy
AIBN	Azobisisobutyronitrile
$AlCl_3$	Aluminum chloride
AM	Additive manufacturing
BF_3	Boron trifluoride
BPA	Bisphenol A
BPO	Benzoyl peroxide
CAD	Computer-aided design
CCl_3COOH	Trichloroacetic acid
CCl_4	Carbon tetrachloride
CFRP	Carbon fiber-reinforced polymer
CFRPs	Carbon fiber-reinforced composites
CMC	Carboxymethyl cellulose
$COCl_2$	Carbonyl chloride (phosgene)
CPE	Chlorinated polyethylene
CPG	Controlled pore glass
DCP	Dicumyl peroxide
DDM	n-Dodecyl mercaptan
DFM	Design for Manufacturability
DIVEMA	Divinyl Ether-Maleic Anhydride
DMA	Dynamic mechanical analysis
DMDO	Dimercapto-3,6-dioxaoctane
DNA	Deoxyribonucleic acid
DP	Degree of polymerization
DSC	Differential scanning calorimetry
ELPs	Electroluminescent polymers
EMA	Ethylene-maleic anhydride
EOL	End-of-life solutions
EPM	Ethylene-propylene monomer
EPR	Extended Producer Responsibility
EPS	Expanded polystyrene
FDM	Fused deposition modeling
FRP	Free radical polymerization
FTIR	Fourier Transform Infrared Spectroscopy
GFC	Gel filtration chromatography
GPC	Gel permeation chromatography
GPPS	General-purpose polystyrene
H_2SO_4	Sulfuric acid
H_3PO_4	Phosphoric acid
HCl	Hydrogen chloride
HDPE	High-density polyethylene
HDXLPE	High-density cross-linked polyethylene
HIPS	High-impact polystyrene
HM	High methoxy
HPPs	High-performance polymers

https://doi.org/10.1515/9783111585734-010

LCA	Life cycle assessment
LDPE	Low-density polyethylene
LLDPE	Linear low-density polyethylene
LM	Low methoxy
MDI	Methylene diisocyanate
MDPE	Medium-density polyethylene
MF	Melamine-formaldehyde
MWD	Molecular weight distribution
NASA	National Aeronautics and Space Administration
NBR	Nitrile butadiene rubber
NMR	Nuclear Magnetic Resonance
NOM	n-**O**ctyl mercaptan
NR	Nitrile rubber
NSA	2-Naphthalene sulfonic acid
OLED	Organic light emitting diode
PA	Polyamide
PAA	Polyacrylic acid
PAM or pAAM	Polyacrylamide
PBAT	Polybutylene adipate terephthalate
PBS	Polybutylene succinate
PC	Polycarbonate
PCL	Polycaprolactone
PDADMAC	Poly(diallyldimethylammonium chloride)
PDI	Polydispersity index
PDMS	Polydi**m**ethyl**s**iloxane
PE	Polyethylene
PEEK	Polyether ether ketone
PEG	Polyethylene glycol
PEI	Poly(ethyleneimine)
PEI	Polyetherimide
PES	Polyethersulfone
PET	Polyethylene terephthalate
PEVA	Polyethylene-vinyl acetate
PHA	Polyhydroxyalkanoates
PI	Polyimide
PIB	Polyisobutylene
PIC	Photo-induced cross-linking
PLA	Polylactic acid
PLGA	Polylactic-co-glycolic acid
PMMA	Poly**m**ethyl **m**ethacrylate
PNIPAM	Poly(N-isopropylacrylamide)
POM	Polyoxymethylene
PP	Polypropylene
PPE	Polyphenylene ether **or** poly(phenylene ethynylene)
PPO	Polyphenylene oxide
PPS	Polyphenylene sulfide
PPV	Poly(p-phenylene vinylene)
PS	Polystyrene
PS-g-PMMA	Polystyrene-graft-poly(methyl methacrylate)

PTFE	Polytetrafluoroethylene (Teflon)
PUA	Polyurea
PUR	Polyurethane
PVAc	Polyvinyl acetate
PVC	polyvinyl chloride
PVDF	Polyvinylidene fluoride
PVP	Polyvinylpyrrolidone
RNA	Ribonucleic acid
ROP	Ring-opening polymerization
rPET	Recycled polyethylene terephthalate
SAN	Styrene-acrylonitrile
SBR	Styrene-butadiene rubber
SEC	Size exclusion chromatography
SEM	Scanning Electron Microscopy
SIN	Simultaneous interpenetrating network
SMA	Styrene-maleic anhydride
$SnCl_4$	Stannic chloride
TEM	Transmission Electron Microscopy
Tg	Glass transition temperature
TGA	Thermogravimetric Analysis
$TiCl_4$	Titanium tetrachloride
UHMWPE	Ultra-high molecular weight polyethylene
ULMWPE	Ultra-low molecular weight polyethylene
UX	User experience
WTE	Waste-to-energy
XLPE (PE**X**)	**C**rosslinked polyethylene
XRD	X-ray diffraction
XRF	X-ray fluorescence

Resources and Further Readings

Books

[1] Elzagheid, M. Macromolecular Chemistry: Natural & Synthetic Polymers, Walter de Gruyter GmbH & Co KG, **2021**, ISBN: 9783110762754s.

[2] Elzagheid, M. Organic Chemistry: 25 Must-Know Classes of Organic Compounds, Walter de Gruyter GmbH & Co KG, 2nd edition, **2024**, ISBN: 9783111381992.

[3] Elzagheid, M. Biomacromolecules: Carbohydrates, Lipids, Proteins and Nucleic Acids, Walter de Gruyter GmbH & Co KG, 2nd edition, **2024**, ISBN: 9783111582986.

[4] Malcolm P. Stevens. Polymer Chemistry: An Introduction, Oxford University Press; 3rd edition, **1998**, ISBN: 9780195124446.

[5] Ravve A. Principles of Polymer Chemistry, Springer; 3rd edition, **2016**, ISBN: 9781493950386.

[6] Fred J. Davis. Polymer Chemistry: A Practical Approach, Oxford University Press; 1st edition, **2004**, ISBN: 9780198503095

[7] Barbara H. Stuart. Polymer Analysis, Wiley, 1st edition, **2002**, ISBN: 9780471813637.

[8] Robert O. Ebewele. Polymer Science and Technology. CRC Press, 1st edition, **2000**, ISBN: 9780849389399.

[9] Robert J. Young, and Peter A. Lovell. Introduction to Polymers, Routledge; 3rd edition, **2011**, ISBN: 9780849339295.

[10] Joel R. Fried. Polymer Science and Technology, Pearson, 3rd edition, **2014**, ISBN: 9780137039555.

[11] Bahadur P., and Sastry N.V. Principles of Polymer Science, Alpha Science; 2nd edition, **2005**, ISBN: 9781842652466.

[12] Timothy P. Lodge, and Paul C. Hiemenz. Polymer Chemistry: International Student Edition, CRC Press, 3rd edition, **2020**, ISBN: 9781466581647.

[13] Charles E. Carraher Jr. Introduction to Polymer Chemistry, CRC Press; 4th edition, **2017**, ISBN: 9781498737616.

[14] Charles E. Carraher Jr. Seymour/Carraher's Polymer Chemistry, CRC Press; 7th edition, **2008**, ISBN: 9781420051025.

[15] George Odian. Principles of Polymerization, Wiley-Interscience; 4th edition, **2004**, ISBN: 9780471274001.

[16] Manas Chanda. Introduction to Polymer Science and Chemistry. Taylor & Francis, Boca Raton, FL., 1st edition, **2006**, eBook ISBN: 9780429128721.

[17] A. Brent Strong. Plastics: Materials and Processing, Pearson; 3rd edition, **2005**, ISBN: 9780131145580.

[18] Solomons TWG, Fryhle CB, Snyder SA. Organic Chemistry, John Wiley & Sons, **2016**, ISBN: 9781118875766.

[19] Wade, L. G. Organic Chemistry, Pearson, **2014**, ISBN: 9781292021652.

[20] McMurry, J. Organic Chemistry, Brooks Cole, **2011**, ISBN: 9780840054449.

[21] Vollhardt, P., and Schore, N. Organic Chemistry "Structure and Function", Freeman and Company, **2014**, ISBN: 139871464120275.

[22] Brown, W. H., Iverson, B. L., Anslyn, E. V., and Foote, C. S. Organic Chemistry, Wadsworth, Cengage Learning, **2014**, ISBN: 139871285052816.

[23] Brown, W. H., and Poon, T. Organic Chemistry, John Wiley & Sons, Inc., **2016**, ISBN: 9781118-875803.

[24] Solomons, T. W. G., Fryhle, C. B., Snyder, S. A. Organic Chemistry, John Wiley & Sons, Inc, **2013**, ISBN: 9781118147399.

[25] Allcock H.R., Frederick Walter Lampe. Contemporary Polymer Chemistry, Pearson College Div; Subsequent edition, **2003**, ISBN: 9780130650566.

https://doi.org/10.1515/9783111585734-011

[26] Viney, C. Techniques for Polymer Organization and Morphology Characterization. Wiley, Hoboken, NJ., 1st edition, **2003**, ISBN: 9780471490708.

[27] Campbell, D., Pethrick, R., and White, J.R. Polymer Characterization: Physical Techniques, 2nd edition, Stanley Thornes, Cheltenham, UK., **2000**, ISBN: 9780748740055.

[28] Groenewoud, W. M., Characterisation of Polymers by Thermal Analysis, Elsevier Science; 1st edition, **2001**, ISBN: 9780444506047.

[29] Vishu Shah. Handbook of Plastics Testing and Failure Analysis, John Wiley & Sons, Inc., 3rd Edition, **2006**, ISBN: 9780470-100424.

[30] Shakeel A., Osmani R.A.M. Handbook of Biodegradable Polymers: Applications in Biomedical Sciences, Industry, and the Environment, Jenny Stanford Publishing; 1st edition, **2024**, ISBN: 9789814968843.

Internet Resources

Designed Monomers and Polymers: Polymer history. https://www.tandfonline.com/doi/pdf/10.1163/156855508X292383. Accessed on September 7, 2024.

Where Does Rubber Come From? https://www.holzrubber.com/education/history-of-rubber/. Accessed on September 14, 2024.

The First Synthetic Plastic. https://www.sciencehistory.org/education/classroom-activities/role-playing-games/case-of-plastics/history-and-future-of-plastics/. Accessed on October 28, 2024.

Polystyrene. https://en.wikipedia.org/wiki/Polystyrene. Accessed on October 5, 2024.

Bioplastics Boom. https://www.americanscientist.org/article/bioplastics-boom. Accessed on October 5, 2024.

Biodegradable Polymers Boom: Tracking Global Market Growth in Tons. https://www.bccresearch.com/pressroom/pls/biodegradable-polymers-boom?srsltid=AfmBOoqgf6EpYY5grHMEa0QCKSF8RuZRiGgmLWHRdaDcNRBdgFEbGzwN. Accessed on October 5, 2024.

Engineering Plastic. https://www.sciencedirect.com/topics/materials-science/engineering-plastic. Accessed on October 12, 2024.

Engineering plastics. https://en.wikipedia.org/wiki/Engineering_plastic#:~:text=Engineering%20plastics%20have%20gradually%20replaced,manufacture%2C%20especially%20in%20complicated%20shapes. Accessed on October 12, 2024.

Timeline of Plastic Development. https://en.wikipedia.org/wiki/Timeline_of_plastic_development. Accessed on October 19, 2024.

Polymers by Functional Groups. https://www.nanosoftpolymers.com/product-category/nsp-polymers-by-functional-groups/#:~:text=The%20functional%20groups%20include%20but,DBCO)%2C%20folate%2C%20etc. Accessed on October 19, 2024.

Polymer. https://en.wikipedia.org/wiki/Polymer. Accessed on October 19, 2024.

A Brief Guide to Polymer Nomenclature. https://www.rsc.org/globalassets/05-journals-books-databases/our-journals/polymer-chemistry/brief-guide-to-polymer-nomenclature.pdf. Accessed on October 26, 2024.

Choosing Initiators (for anionic polymerization). https://ocw.mit.edu/courses/10-569-synthesis-of-polymers-fall-2006/e51ca6db1ae52f5f45c48906afe1bd22_lec20_10302006.pdf. Accessed on November 2, 2024.

Ring-Opening Polymerization. https://en.wikipedia.org/wiki/Ring-opening_polymerization. Accessed on November 12, 2024.

Ring-Opening Polymerization. https://www.degruyter.com/document/doi/10.1515/cti-2020-0028/html. Accessed on November 12, 2024.

Recent Advances in the Ring-Opening Polymerization of Sulfur-Containing Monomers. https://pubs.rsc.org/en/content/articlehtml/2022/py/d2py00831a. Accessed on November 12, 2024.

Condensation Polymer. https://en.wikipedia.org/wiki/Condensation_polymer. Accessed on November 17, 2024.

Condensation Polymers (A-Level Chemistry). https://studymind.co.uk/notes/condensation-polymers/. Accessed on November 17, 2024.

Condensation Polymers. https://scienceready.com.au/pages/condensation-polymers. Accessed on November 17, 2024.

Condensation Polymers: Polyamides. https://wisc.pb.unizin.org/chem109fall2021ver02/chapter/polyamides/. Accessed on November 17, 2024.

The Chemistry of Polyurethanes. https://www.l-i.co.uk/knowledge-centre/the-chemistry-of-polyurethanes/. Accessed on November 23, 2024.

Polycarbonates. https://www.essentialchemicalindustry.org/polymers/polycarbonates.html. Accessed on November 23, 2024.

Copolymer. https://byjus.com/chemistry/copolymer/. Accessed on November 27, 2024.

Random Copolymerization Strategy for Non-halogenated Solvent-Processed All-Polymer Solar Cells with a High Efficiency of over 17%. https://pubs.rsc.org/en/content/articlelanding/2022/ee/d2ee01996e. Accessed on November 27, 2024.

Random Copolymerization of Macromonomers as a Versatile Strategy to Synthesize Mixed-Graft Block Copolymers. https://onlinelibrary.wiley.com/doi/abs/10.1002/pol.20210407. Accessed on November 27, 2024.

Random Copolymer. https://www.sciencedirect.com/topics/engineering/random-copolymer. Accessed on November 27, 2024.

Alternating Copolymer. https://www.sciencedirect.com/topics/chemistry/alternating-copolymer. Accessed on November 28, 2024.

Graft Copolymer. https://www.sciencedirect.com/topics/materials-science/graft-copolymer#:~:text=Examples%20of%20graft%20copolymers%20are,%2Dpoly(ethylene%20oxide). Accessed on November 28, 2024.

Block Copolymer. https://www.sciencedirect.com/topics/earth-and-planetary-sciences/block-copolymer. Accessed on November 28, 2024.

Free Radical Polymerization Kinetics. https://uvebtech.com/articles/2023/free-radical-polymerization-kinetics/. Accessed on November 30, 2024.

Molar Mass Distribution. https://www.sciencedirect.com/topics/chemistry/molar-mass-distribution. Accessed on December 05, 2024.

Distribution of Molecular Weights. https://www.petrocuyo.com/en/technology/polypropylene/molecular-weights. Accessed on December 05, 2024.

Functionalization and Modification of Polyethylene Terephthalate Polymer by AgCl Nanoparticles under Ultrasound Irradiation as Bactericidal. https://pubs.acs.org/doi/10.1021/acsomega.1c07082. Accessed on December 07, 2024.

Modification of Polymers. https://www.sciencedirect.com/science/article/abs/pii/B9780128168066000054. Accessed on December 07, 2024.

Polymer Blends. https://kruger.industries/polymer-blends/. Accessed on December 07, 2024.

New Method Based on Direct Analysis in Real-Time Coupled with Time-of-Flight Mass Spectrometry (DART-ToF-MS) for Investigation of the Miscibility of Polymer Blends. https://www.mdpi.com/2073-4360/14/9/1644. Accessed on December 07, 2024.

Cross-Linking. https://www.biologyonline.com/dictionary/cross-linking. Accessed on December 15, 2024.

Polymer Blend Techniques for Designing Carbon Materials. https://www.sciencedirect.com/science/article/abs/pii/S0008622316309095. Accessed on December 15, 2024.

Polymer Additives and Blends. https://www.slideshare.net/slideshow/polymer-additives-and-blends/236322574. Accessed on December 15, 2024.

Latex Blending. https://www.sciencedirect.com/topics/engineering/latex-blending#:~:text=Latex%20blend ing%20is%20similar%20to,10). Accessed on December 15, 2024.

Polymer Crosslinking. https://www.sciencedirect.com/topics/materials-science/polymer-crosslinking. Accessed on December 18, 2024.

Ionic Crosslinking. https://www.sciencedirect.com/topics/engineering/ionic-crosslinking. Accessed on December 18, 2024.

Chemical Crosslinking. https://www.sciencedirect.com/topics/engineering/chemical-crosslinking. Accessed on December 20, 2024.

Arkema Organic Peroxides Crosslinking. https://www.rbhltd.com/market-news/arkema-organic-peroxides-in-crosslinking/. Accessed on December 20, 2024.

Ionic Crosslinking. https://www.sciencedirect.com/topics/engineering/ionic-crosslinking#:~:text=For%20ex ample%2C%20adding%20divalent%20cations,thermal%20denaturation%20through%20hydrophobic %20interactions. Accessed on December 20, 2024.

Alginate. https://theory.labster.com/alginate/. Accessed on December 20, 2024.

Ionic Crosslinking. https://theory.labster.com/ionic-crosslinking/. Accessed on December 20, 2024.

Photo-induced covalent cross-linking for the analysis of biomolecular interactions. https://pubs.rsc.org/ en/content/articlelanding/2013/cs/c3cs35459h. Accessed on December 22, 2024.

Photochemistry and Photoinduced Chemical Crosslinking Activity of Acrylated Prepolymers by Several Commercial Type I Far UV Photoinitiators. https://www.sciencedirect.com/science/article/abs/pii/ S0141391099000038. Accessed on December 22, 2024.

The Application of Ultraviolet-Induced Photo-Crosslinking in Edible Film Preparation and Its Implication in Food Safety. https://www.sciencedirect.com/science/article/abs/pii/S0023643820307805. Accessed on December 22, 2024.

How to Enhance Polymer Properties Through Crosslinking? https://polymer-additives.specialchem.com/ tech-library/article/crosslinking. Accessed on December 22, 2024.

Crosslinked Thermoplastics. https://www.sciencedirect.com/science/article/abs/pii/B9781455731077000178. Accessed on January 02, 2025.

Crosslinked Thermoset Resins and Methods Thereof. https://patents.google.com/patent/ US20150315249A1/en. Accessed on January 02, 2025.

Dynamic Aliphatic Polyester Elastomers Crosslinked with Aliphatic Dianhydrides. https://pubs.acs.org/doi/ 10.1021/acspolymersau.3c00004. Accessed on January 02, 2025.

Cross-Linked Hydrogel for Pharmaceutical Applications: A Review. https://pmc.ncbi.nlm.nih.gov/articles/ PMC5788207/. Accessed on January 02, 2025.

Biopolymers: Manufacturing's Latest Green Hero? https://www.cas.org/resources/cas-insights/biopoly mers-manufacturings-latest-green-hero. Accessed on January 04, 2025.

Biopolymers. https://link.springer.com/referenceworkentry/10.1007/978-981-19-0710-4_1. Accessed on January 04, 2025.

Biopolymer. https://www.sciencedirect.com/topics/medicine-and-dentistry/biopolymer. Accessed on January 04, 2025.

Basics of Green Chemistry. https://www.epa.gov/greenchemistry/basics-green-chemistry#:~:text=Green% 20chemistry%20is%20the%20design,%2C%20use%2C%20and%20ultimate%20disposal. Accessed on January 04, 2025.

Principles of Green Chemistry. https://greenchemistry.yale.edu/about/principles-green-chemistry. Accessed on January 05, 2025.

Polyethylene. https://byjus.com/chemistry/polyethylene/. Accessed on January 05, 2025.

Polymer Crystallization. https://www.ipfdd.de/en/research/institute-theory-of-polymers/fields-of-research /theory-of-soft-matter-and-polymer-physics/polymer-crystallization/. Accessed on January 06, 2025.

Unusual Crystalline Morphology of Poly Aryl Ether Ketones (PAEKs). https://pubs.rsc.org/en/content/arti clehtml/2016/ra/c5ra17110e. Accessed on January 06, 2025.

Liquid Polysulfide Polymers for Chemical- and Solvent-Resistant Sealants. https://www.adhesivesmag. com/articles/96384-liquid-polysulfide-polymers-for-chemical–and-solvent-resistant-sealants. Accessed on January 07, 2025.

Correlations Between Precursor Molecular Weight and Dynamic Mechanical Properties of Polyborosiloxane (PBS). https://onlinelibrary.wiley.com/doi/abs/10.1002/mame.202100360. Accessed on January 07, 2025.

Preparation, Structure and Properties of Hybrid Materials Based on Geopolymers and Polysiloxanes. https://www.sciencedirect.com/science/article/abs/pii/S0264127515302562. Accessed on January 08, 2025.

Polymeric Support. https://www.sciencedirect.com/topics/chemistry/polymeric-support#:~:text=The% 20most%20commonly%20used%20polymeric,(Wang%2Dtype%20resin). Accessed on January 12, 2025.

Peptide Synthesis. https://www.biotech.iastate.edu/protein/resources/peptide-synthesis/#:~:text=An%20in soluble%20polymer%20support%20(resin,in%20solid%20phase%20peptide%20synthesis. Accessed on January 12, 2025.

Solid-Phase Oligonucleotide Synthesis. https://atdbio.com/nucleic-acids-book/Solid-phase-oligonucleotide-synthesis. January 14, 2025.

High Performance Polymer. https://www.sciencedirect.com/topics/materials-science/high-performance-polymer. Accessed on January 14, 2025.

Polymer Blend. https://www.sciencedirect.com/topics/materials-science/polymer-blend#:~:text=The% 20morphology%20of%20the%20immiscible,interfacial%20tension%20between%20the%20polymers. Accessed on January 18, 2025.

The Age of Composite Materials: History, Classification & Applications. https://fabheads.com/blogs/the-age-of-composite-materials-history-classification-applications/. Accessed on January 18, 2025.

Nanocomposite Materials. https://www.intechopen.com/chapters/72636. Accessed on January 20, 2025.

Head-to-Head Polymers. https://www.sciencedirect.com/science/article/pii/S0079670099000325#:~:text= Definition%20of%20head%20to%20tail,1%5D%2C%20%5B2%5D. Accessed on January 26, 2025.

Stereoisomerism. https://www.open.edu/openlearn/science-maths-technology/chemistry/introduction-polymers/content-section-2.3.4. Accessed on January 26, 2025.

Morphology of Polymer Blends. https://link.springer.com/referenceworkentry/10.1007/0-306-48244-4_8. Accessed on January 26, 2025.

Fourier Transform Infrared Spectroscopy Analysis of Polymers and Plastics. https://www.intertek.com/poly mers-plastics/analysis-ftir/. Accessed on January 28, 2025.

Fourier Transform Infrared Spectrometer. https://www.sciencedirect.com/topics/engineering/fourier-transform-infrared-spectrometer. Accessed on January 28, 2025.

Gel Permeation Chromatography. https://www.sciencedirect.com/topics/materials-science/gel-permeation -chromatography#:~:text=Gel%20permeation%20chromatography%20(GPC)%20or,correlated% 20directly%20to%20molecular%20mass. Accessed on February 02, 2025.

Gel Permeation Chromatography/ Size Exclusion Chromatography. https://itn-snal.net/2014/11/gel-permeation-chromatography-size-exclusion-chromatography/. Accessed on February 02, 2025.

Viscosity of Polymers in Solution. https://www.aimplas.net/blog/viscosity-polymers-solution/. Accessed on February 03, 2025.

What Is the Difference Between Ostwald and Ubbelohde Viscometers. https://pediaa.com/what-is-the-difference-between-ostwald-and-ubbelohde-viscometers/. Accessed on February 03, 2025.

How to Measure Viscosity. https://wiki.anton-paar.com/en-ae/how-to-measure-viscosity/. Accessed on February 04, 2025.

Differential Scanning Calorimetry. https://www.sciencedirect.com/topics/materials-science/differential-scanning-calorimetry#:~:text=Differential%20scanning%20calorimetry%20(DSC)%20monitors,as%20a %20function%20of%20temperature. Accessed on February 04, 2025.

Atomic Force Microscopy. https://www.sciencedirect.com/topics/chemistry/atomic-force-microscopy. Accessed on February 05, 2025.

Polymer testing. https://www.polymertesting.com.au/services/. Accessed on February 05, 2025.

Chemical Resistance of Polymers. https://omnexus.specialchem.com/polymer-property/chemical-resistance-polymers. Accessed on February 05, 2025.

Performance Tests to Ensure Suitability for UV Resistance. https://geosyntheticsmagazine.com/2022/07/08/performance-tests-to-ensure-suitability-for-uv-resistance/. Accessed on February 06, 2025.

Injection Molding: The Complete Guide to Precision Plastic Manufacturing. https://omnexus.specialchem.com/selection-guide/injection-molding. Accessed on February 06, 2025.

What Is Thermoforming? https://radiofrequencywelding.com/what-is-thermoforming-and-how-can-the-process-be-used-in-product-manufacturing/. Accessed on February 06, 2025.

Polymer Casting. https://www.modulusmetal.com/services/molding/polymer-casting/. Accessed on February 06, 2025.

Polymer Casting. https://engineeringtechnology.org/manufacturing/casting-processes/polymer-casting/. Accessed on February 07, 2025.

Recent Trends of Foaming in Polymer Processing: A Review. https://pmc.ncbi.nlm.nih.gov/articles/PMC6631771/. Accessed on February 07, 2025.

3 Common Types of Lamination Process in Packaging. https://kdwpack.com/lamination-process-in-packaging/. Accessed on February 07, 2025.

Calendering Process Basics: Knowing Its Principles and Application. https://leadrp.net/blog/calendering-process-basics-knowing-its-principles-and-application/. Accessed on February 10, 2025.

Different types of Spinning Techniques. https://vnpolyfiber.com/different-types-of-spinning-techniques/. Accessed on February 10, 2025.

Chemical Fiber Spinning Methods for Filament Yarns. https://www.textileblog.com/chemical-fiber-spinning-methods-for-filament-yarns/. Accessed on February 10, 2025.

Biodegradable Polymers. https://pmc.ncbi.nlm.nih.gov/articles/PMC5445709/. Accessed on February 12, 2025.

Conductive Polymer. https://www.sciencedirect.com/topics/engineering/conductive-polymer#:~:text=Conductive%20polymers%20are%20a%20subset,polyaniline%2C%20polypyrrole%2C%20and%20polythiophene. Accessed on February 12, 2025.

Biomedical Polymers. https://www.sciencedirect.com/topics/materials-science/biomedical-polymer. Accessed on February 14, 2025.

Support for Solid Phase Synthesis of Oligonucleotides by Radiation-Induced Gas Phase Grafting of Styrene onto Polytetrafluoroethylene. https://www.sciencedirect.com/science/article/abs/pii/1359019789902476. Accessed on February 14, 2025.

High Performance Polymer. https://www.sciencedirect.com/topics/materials-science/high-performance-polymer. Accessed on February 15, 2025.

Smart Polymers. https://en.wikipedia.org/wiki/Smart_polymer. Accessed on February 15, 2025.

Smart Polymer. https://www.sciencedirect.com/topics/engineering/smart-polymer. Accessed on February 15, 2025.

Plastic Resins and Compounds for Sustainable Solutions. https://www.entecpolymers.com/resources/news/plastic-resins-and-compounds-for-sustainable-solutions. Accessed on February 17, 2025.

What Is Recycled Polyethylene Terephthalate (rPET). https://www.reusethisbag.com/articles/what-is-recycled-polyethylene-terephtalate. Accessed on February 17, 2025.

Water Soluble Polymers for Pharmaceutical Applications. https://www.mdpi.com/2073-4360/3/4/1972. Accessed on February 17, 2025.

Electroluminescent Polymers. https://link.springer.com/referenceworkentry/10.1007/978-3-642-36199-9_104-2. Accessed on February 19, 2025.

Electroluminescent Polymers. https://www.sciencedirect.com/science/article/abs/pii/S0079670002001405. Accessed on February 19, 2025.

Exploring Microgel Adsorption: Synthesis, Classification, and Pollutant Removal Dynamics. https://pubs.rsc.org/en/content/articlehtml/2024/ra/d4ra00563e. Accessed on February 22, 2025.

Some Examples of Polymer Selection. https://link.springer.com/chapter/10.1007/978-94-009-4101-4_3. Accessed on February 25, 2025.

Polymer Products: Design, Materials and Processing. https://link.springer.com/book/10.1007/978-94-009-4101-4. Accessed on February 25, 2025.

General Design Principles for DuPont Engineering Polymers. https://www.distrupol.com/General_Design_Principles_for_Engineering_Polymers.pdf. Accessed on February 25, 2025.

Design for Manufacturability in Polymer Optics. https://gandh.com/news-and-resources/design-for-manufacturability. Accessed on February 28, 2025.

Design for Manufacturability and Its Importance. https://www.polymershapes.com/dfm-plastics/. Accessed on February 28, 2025.

The Aesthetic of Interaction with Materials for Design: The Bioplastics' Identity. https://dl.acm.org/doi/10.1145/2347504.2347540. Accessed on March 1, 2025.

Polymer Chemistry Unit 11 – Industrial Polymer Applications. https://library.fiveable.me/polymer-chemistry/unit-11. Accessed on March 1, 2025.

Automotive Polymers. https://www.elastomer.kuraray.com/applications/automotive-polymers/. Accessed on March 1, 2025.

Aerospace. https://nationalpolymer.com/aerospace/. Accessed on March 1, 2025.

Custom Organic Electronics Out of the Printer. https://als.lbl.gov/custom-organic-electronics-out-of-the-printer/. Accessed on March 1, 2025.

Medical Industry. https://www.cdiproducts.com/solutions/markets/medical. Accessed on March 1, 2025.

The Importance of Plastic Polymers in Packaging. http://www.packcon.org/index.php/en/articles/113-2017new/200-the-importance-of-plastic-polymers-in-packaging. Accessed on March 1, 2025.

The Use of Polymer Materials in the Textile Industry https://www.iranpetroleum.co/the-use-of-polymer-materials-in-the-textile-industry/. Accessed on March 1, 2025.

What's the Difference Between Upcycling and Downcycling? https://www.roadrunnerwm.com/blog/difference-between-upcycling-and-downcycling. Accessed on March 4, 2025.

Recycling 101 – What Do the Plastic Codes Mean? https://methodrecycling.com/world/journal/recycling-101-what-do-the-plastic-codes-mean. Accessed on March 4, 2025.

Recycling Codes. https://en.wikipedia.org/wiki/Recycling_codes. Accessed on March 4, 2025.

1, 2, 3, 4, 5, 6, 7: Plastics Recycling By the Numbers. https://millerrecycling.com/plastics-recycling-numbers/. Accessed on March 4, 2025.

Types of Plastic and Their Recycling Codes. https://dpw.lacounty.gov/epd/SBR/pdfs/TypesOfPlastic.pdf. Accessed on March 4, 2025.

Polymer and Its Effect on Environment. https://www.sciencedirect.com/science/article/abs/pii/S0019452222004836. Accessed on March 8, 2025.

Environmental Problems Caused by Polymers. https://www.azocleantech.com/article.aspx?ArticleID=1189. Accessed on March 8, 2025.

Environmental Impact of Polymers. https://www.wiley.com/en-br/Environmental±Impact±of±Polymers-p-9781848216211. Accessed on March 8, 2025.

Degradation of Polymer Materials in the Environment and Its Impact on the Health of Experimental Animals: A Review. https://www.mdpi.com/2073-4360/16/19/2807. Accessed on March 8, 2025.

Appendices

Appendix A (Trade Names for Common Polymers)

Polyolefins

Polyethylene (PE) – Marlex®, Hostalen®, Alathon®
High-Density Polyethylene (HDPE) – Lupolen®, Alathon®, Hostalen®
Low-Density Polyethylene (LDPE) – Dowlex®, Lupolen®, Sclair®
Linear Low-Density Polyethylene (LLDPE) – ExxonMobil Exact®, Dow Engage®
Polypropylene (PP) – Moplen®, Profax®, Hostacom®
Ethylene Vinyl Acetate (EVA) – Elvax®, Levapren®, Vistamaxx®

Polyesters

Polyethylene Terephthalate (PET) – Mylar®, Dacron®, Arnite®
Polybutylene Terephthalate (PBT) – Ultradur®, Valox®, Crastin®
Polytrimethylene Terephthalate (PTT) – Sorona®, Corterra®

Polyamides (Nylons)

Nylon 6 (PA6) – Zytel®, Ultramid®, Capron®
Nylon 6,6 (PA6,6) – Zytel®, Stanyl®, Vydyne®
Nylon 11 (PA11) – Rilsan®
Nylon 12 (PA12) – Vestamid®, Grilamid®

Fluoropolymers

Polytetrafluoroethylene (PTFE) – Teflon®, Fluon®, Hostaflon®
Polyvinylidene Fluoride (PVDF) – Kynar®, Solef®
Ethylene Tetrafluoroethylene (ETFE) – Tefzel®
Fluorinated Ethylene Propylene (FEP) – Teflon FEP®

Polycarbonates and Acrylics

Polycarbonate (PC) – Lexan®, Makrolon®, Calibre®
Polymethyl Methacrylate (PMMA) – Plexiglas®, Lucite®, Acrylite®

Styrenic Polymers

Polystyrene (PS) – Styron®, Styrolux®, Edistir®
High-Impact Polystyrene (HIPS) – Styron®, Edistir®
Acrylonitrile Butadiene Styrene (ABS) – Cycolac®, Terluran®, Lustran®
Styrene-Acrylonitrile (SAN) – Lustran SAN®, Tyril®

High-Performance Polymers

Polyetheretherketone (PEEK) – Victrex®, KetaSpire®
Polyphenylene Sulfide (PPS) – Ryton®, Fortron®
Liquid Crystal Polymer (LCP) – Vectra®, Xydar®
Polyimide (PI) – Kapton®, Vespel®
Polybenzimidazole (PBI) – Celazole®
Polyetherimide (PEI) – Ultem®

---

https://doi.org/10.1515/9783111585734-012

(continued)

Specialty and Biodegradable Polymers
Polylactic Acid (PLA) – Ingeo®, Revode®, NatureWorks®
Polyhydroxyalkanoates (PHA) – Mirel®, Nodax®

Elastomers and Rubbers
Polyurethane (PU) – Desmopan®, Elastollan®, Pellethane®
Ethylene Propylene Diene Monomer (EPDM) – Nordel®, Keltan®, Vistalon®
Polybutadiene (BR) – Taktene®, Buna®
Styrene-Butadiene Rubber (SBR) – Buna-S®, Europrene®, Duradene®
Polychloroprene (CR, Neoprene) – Neoprene®, Baypren®

Silicones and Thermosets
Polydimethylsiloxane (PDMS, Silicone) – Silastic®, Dow Corning®
Phenol-Formaldehyde (PF, Bakelite) – Bakelite®, Novolac®
Epoxy Resins (EP) – Araldite®, Epikote®, Epon®
Unsaturated Polyester Resins (UPR) – Atlac®, Palatal®, Viapal®

Conductive and Water-Soluble Polymers
Polyaniline (PANI) – Pani-Tex®
Polyethylene Oxide (PEO, PEG) – Carbowax®, Polyox®
Polypropylene Oxide (PPO) – Noryl®, Santoprene®
Polysulfone (PSU) – Udel®, Ultrason®
Polyetherketoneketone (PEKK) – AvaSpire®, Kepstan®

Appendix B (Amorphous, Crystalline, and Semicrystalline Polymers)

Crystalline Polymers

Polyethylene (PE) – High-Density Polyethylene (HDPE) and Ultra-High Molecular Weight Polyethylene (UHMWPE).
Polypropylene (PP) – Especially isotactic polypropylene (iPP).
Polyethylene Terephthalate (PET).
Polyoxymethylene (POM).
Polybutene-1 (PB-1).
Polytetrafluoroethylene (PTFE) – Known as Teflon.
Polyetheretherketone (PEEK).
Syndiotactic Polystyrene (sPS).
Polyamide (PA) – Such as Nylon 6 and Nylon 6,6.
Polyphenylene Sulfide (PPS).
Low-Density Polyethylene (LDPE).
Ethylene-Vinyl Alcohol (EVOH).
Ethylene-Tetrafluoroethylene (ETFE).
Fluorinated Ethylene Propylene (FEP).
Polyvinylidene Fluoride (PVDF).

(continued)

Polytrimethylene Terephthalate (PTT).
Polylactic Acid (PLA).
Polyethylene Naphthalate (PEN).
Polyamide 11 (PA11).
Polyamide 12 (PA12).

Semi-crystalline Polymers

Polyetherimide (PEI).
Liquid-Crystal Polymers (LCPs).
Polybenzimidazole (PBI).
Polyimide (PI).
Polyaryl Ether Ketone (PAEK).
Low-Density Polyethylene (LDPE).
Polypropylene (PP).
Ethylene-Propylene-Diene Monomer (EPDM).
Polytrimethylene Terephthalate (PTT).
Nylon 6 (PA 6).
Nylon 6,6 (PA 6,6).
Nylon 11 (PA 11).
Nylon 12 (PA 12).
Polycarbonate (PC).
Polylactic Acid (PLA).
Polyetheretherketone (PEEK).
Polyphenylene Sulfide (PPS).
Polyetherimide (PEI).
Polyvinylidene Fluoride (PVDF).
Polyoxymethylene (POM).

Amorphous Polymers

Polystyrene (PS).
Acrylonitrile Butadiene Styrene (ABS).
Polycarbonate (PC).
Polyvinyl Chloride (PVC).
Polymethyl Methacrylate (PMMA).
Polyetherimide (PEI).
Polysulfone (PSU).
Polyethersulfone (PES).
Polyimide (PI).
Polyurethane (PU).
Polyisoprene (IR, NR).
Polybutadiene (BR).
Styrene-Butadiene Rubber (SBR).
Polydimethylsiloxane (PDMS).
Polyvinyl Acetate (PVAc).
Polyvinyl Alcohol (PVOH).
Polyhydroxyalkanoates (PHA).

Appendix C (Chemical Structures for Common Polymers Repeating Units)

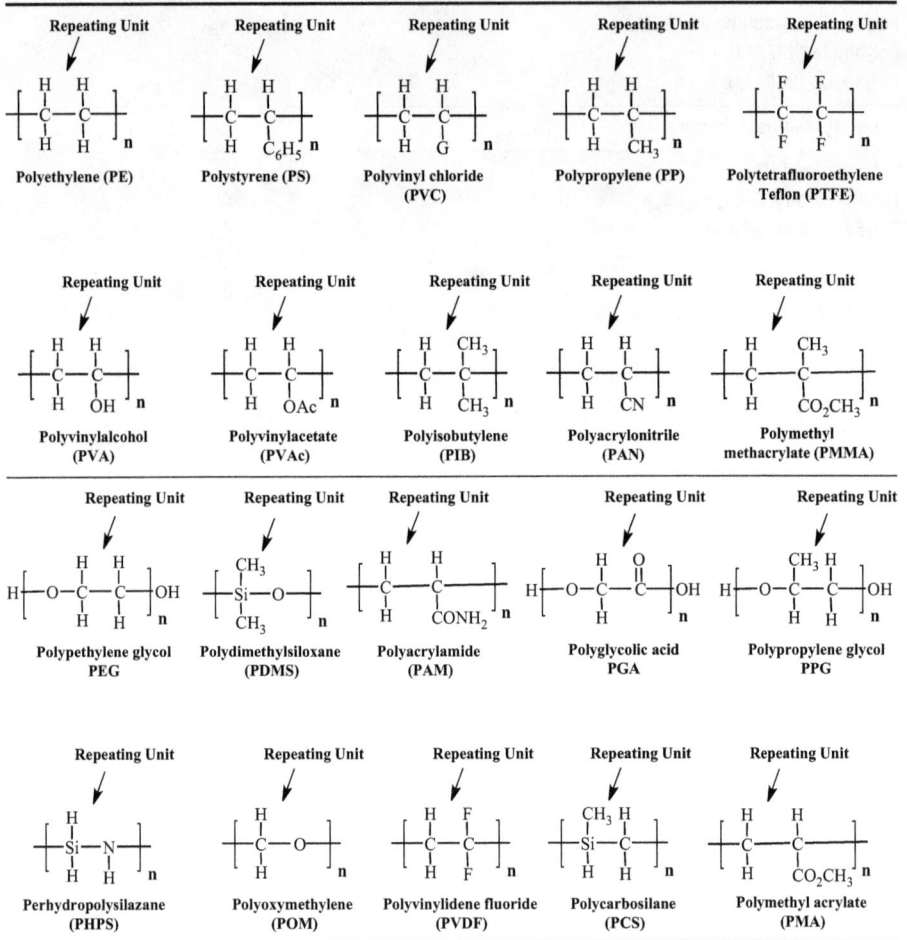

Repeating Unit	Repeating Unit	Repeating Unit	Repeating Unit	Repeating Unit
Polyethylene (PE)	Polystyrene (PS)	Polyvinyl chloride (PVC)	Polypropylene (PP)	Polytetrafluoroethylene Teflon (PTFE)
Polyvinylalcohol (PVA)	Polyvinylacetate (PVAc)	Polyisobutylene (PIB)	Polyacrylonitrile (PAN)	Polymethyl methacrylate (PMMA)
Polypethylene glycol PEG	Polydimethylsiloxane (PDMS)	Polyacrylamide (PAM)	Polyglycolic acid PGA	Polypropylene glycol PPG
Perhydropolysilazane (PHPS)	Polyoxymethylene (POM)	Polyvinylidene fluoride (PVDF)	Polycarbosilane (PCS)	Polymethyl acrylate (PMA)

Appendix D (Polymer Suppliers and Recycling Facilities)

Polymer Suppliers

SABIC (Saudi Basic Industries Corporation)
Riyadh, Saudi Arabia
One of the world's largest petrochemical manufacturers.
Products: Polyethylene, polypropylene, and polyvinyl chloride.

(continued)

Shijiazhuang Honglai Cellulose Co., Ltd.
Hebei, China
Products: Hydroxypropyl Methyl Cellulose, Redispersible Polymer Powder, Hydroxyethyl Cellulose Powder.

Braskem
São Paulo, Brazil
The largest producer of thermoplastic resins in the Americas.
Products: Polyethylene, polypropylene, and polyvinyl chloride.

Borealis AG
Vienna, Austria
Products: Polyolefins, base chemicals, and fertilizers.

NatureWorks
Plymouth, Minnesota, USA
Products: Ingeo™ biopolymer, a polylactic acid derived from renewable resources, used in various applications from packaging to fibers.

ALPLA Group
Hard, Austria
Products: Blow-molded bottles and caps, injection-molded parts, preforms, and tubes, with a significant presence in the global plastic packaging industry.

Evonik Industries
Essen, Germany
Evonik is a specialty chemicals company producing various polymers and additives.
Products: Polymethyl methacrylate (PMMA) and polyamide 12, used in automotive, healthcare, and consumer goods.

BASF
Ludwigshafen, Germany
BASF is the world's largest chemical producer, offering a broad portfolio that includes a variety of polymers.
Products: Polyurethanes, engineering plastics, and biodegradable polymers.

Dow
Midland, Michigan, USA
Dow is a leading chemical company producing a wide range of polymers.
Products: Polyethylene, polypropylene, and polystyrene, serving industries like packaging, automotive, and consumer goods.

LyondellBasell Industries
Houston, Texas, USA
LyondellBasell is one of the largest plastics, chemicals, and refining companies globally.
Products: Polyolefins such as polyethylene and polypropylene.

(continued)

Mitsubishi Chemical Group
Tokyo, Japan
Mitsubishi Chemical is a comprehensive chemical company producing various polymers.
Products: Polyethylene, polypropylene, and specialty resins, serving industries such as automotive, electronics, and healthcare.

Shin-Etsu Chemical
Tokyo, Japan
Shin-Etsu is a leading chemical company.
Products: Polyvinyl chloride (PVC) and silicones, supplying materials for construction, electronics, and automotive industries.

Arkema
Colombes, France
Arkema is a global chemical company offering a range of polymers.
Products: High-performance polyamides and fluoropolymers, catering to industries like automotive, electronics, and construction.

Covestro
Leverkusen, Germany
Products: High-performance polymers, including polyurethanes and polycarbonates, serving sectors like automotive, construction, and electronics.

Recycling Facilities

Ravago
Luxembourg City, Luxembourg
Specialization: Recycling and compounding plastic and elastomeric raw materials.

Veolia Middle East
Riyadh, Saudi Arabia
Specialization: Comprehensive waste management and recycling services.

Saudi Investment Recycling Company (SIRC)
Riyadh, Saudi Arabia
Specialization: Waste management and recycling services.

MBA Polymers
Mauna near Meissen, Germany
Specialization: Recycling waste electrical and electronic equipment (WEEE).

Interzero Holding GmbH & Co. KG
Berlin and Cologne, Germany
Specialization: Environmental services, including recycling and waste management.

(continued)

Wellman International
Kells, Co. Meath, Ireland
Specialization: Recycling PET bottles, polyester fibers.

Pure PET
San Polo di Piave, Italy
Specialization: Recycling PET bottles, PET thermoforms.

Phoenix Technologies
Bowling Green, OH, USA
Specialization: Recycling PET bottles.

Appendix E (Names and Uses of Polymerization Chain Transfer Agents)

Name	Use
n-Dodecyl Mercaptan (DDM)	Often used in radical polymerization to control polymer molecular weight.
2-Mercaptoethanol	Used in water-soluble polymerization systems.
Dithiobenzoic Acid	A key chain transfer agents (CTA) in reversible addition-fragmentation chain transfer (RAFT) polymerization.
Dithiocarbamates (e.g., N,N-Dimethyldithiocarbamate)	Effective in RAFT polymerization.
Ethyl α-Bromoisobutyrate (EBiB)	Used in atom transfer radical polymerization (ATRP) as an initiator and CTA.
Methyl 2-Bromopropionate (MBP)	Used in ATRP to regulate polymerization.
Tris(trimethylsilyl)phosphine	Chain transfer agent in radical polymerization.
n-Octyl Mercaptan (OM)	Used in emulsion and solution polymerization to control molecular weight.
Tert-Dodecyl Mercaptan (TDM)	Effective in rubber and acrylic polymerization.
Thiophenol	Used to modify radical polymerization kinetics.
2-Cyanoprop-2-yl Dithiobenzoate (CPDB)	A widely used RAFT agent for controlled radical polymerization.
Dibenzyl Trithiocarbonate (DBTTC)	Used in RAFT polymerization for controlling polymer molecular weight.

(continued)

Name	Use
4-Cyano-4-(Phenylcarbonothioylthio) pentanoic Acid (CPCTPA)	A highly effective RAFT agent for water-soluble polymers.
Methyl 2-Chloropropionate (MCP)	ATRP initiator and chain transfer agent.
1-Phenylethyl Bromide (PEBr)	Used in ATRP to regulate polymerization kinetics.
2-Bromoisobutyryl Bromide (BiBB)	A halogen-based chain transfer agent for controlled polymerization.
Diethyl Phosphite	Used as a CTA in radical polymerization for modifying polymer structure.
Triphenyl Phosphine (PPh₃)	Used in controlled polymerization and also acts as a ligand in ATRP.
O-Ethyl Xanthic Acid S-Ethyl Ester	Used in RAFT polymerization of vinyl monomers.
Potassium O-Ethyl Xanthate	Used in free radical polymerization and rubber production.

Appendix F (Synthetic Schemes of Common Polymers)

PET

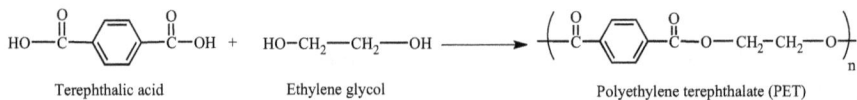

Terephthalic acid   Ethylene glycol   Polyethylene terephthalate (PET)

OR

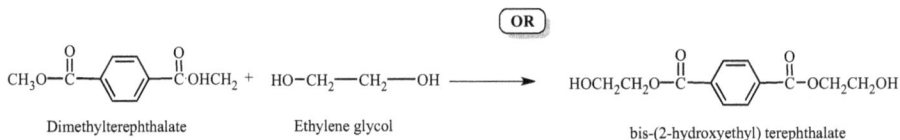

Dimethylterephthalate   Ethylene glycol   bis-(2-hydroxyethyl) terephthalate

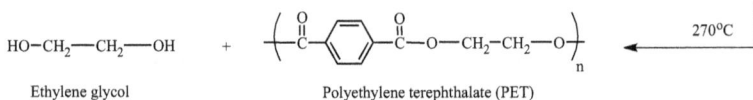

Ethylene glycol   Polyethylene terephthalate (PET)   270°C

(continued)

PUR

Bis(4-isocyanatophenyl) methane   Ethylene glycol

- CO$_2$

Polyurethane (PUR)

PC

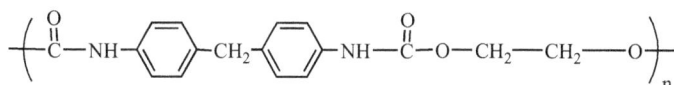

Bisphenol A   Phosgene (Carbonyl chloride)   Polycarbonate (PC)   2 HCl Hydrogen chloride

OR

Diphenylcarbonate   2 Phenol

PA (Nylon 6,6)

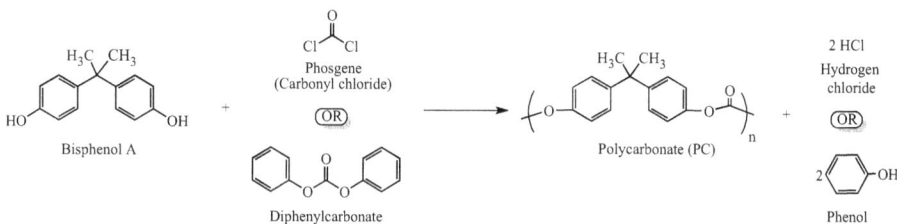

Adipic acid

OR

Adipoyl chloride

Hexamethylenediamine   Nylon 6,6 (Polyamide, PA)

PVC

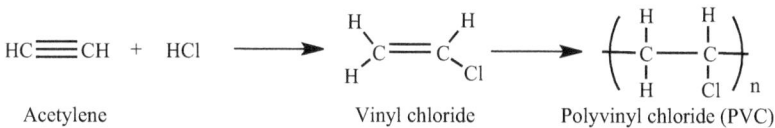

Acetylene   Vinyl chloride   Polyvinyl chloride (PVC)

(continued)

PMMA

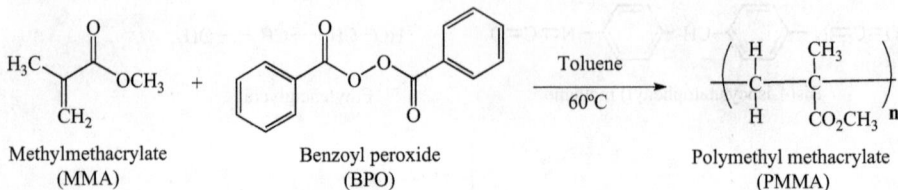

Methylmethacrylate (MMA) + Benzoyl peroxide (BPO), Toluene, 60°C → Polymethyl methacrylate (PMMA)

PAN

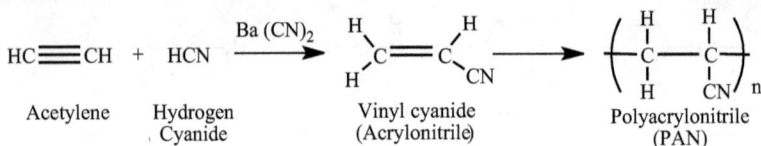

$HC \equiv CH$ (Acetylene) + HCN (Hydrogen Cyanide), Ba(CN)$_2$ → Vinyl cyanide (Acrylonitrile) → Polyacrylonitrile (PAN)

PP

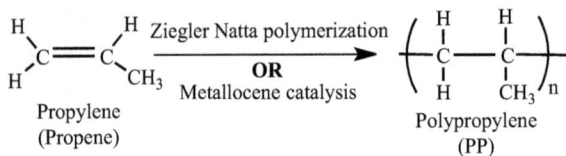

Propylene (Propene), Ziegler Natta polymerization OR Metallocene catalysis → Polypropylene (PP)

PE

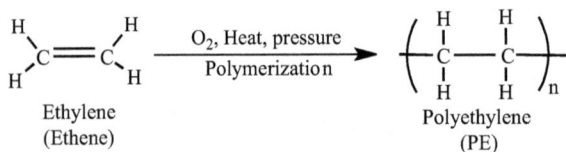

Ethylene (Ethene), O_2, Heat, pressure Polymerization → Polyethylene (PE)

PVAc and PVA

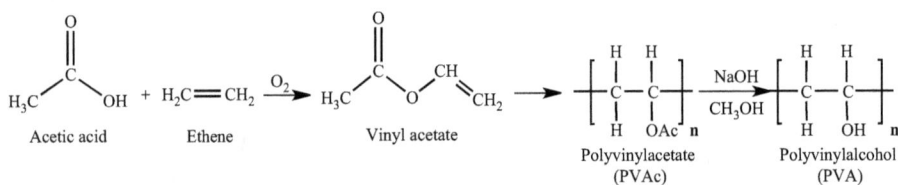

Acetic acid + Ethene, O_2 → Vinyl acetate → Polyvinylacetate (PVAc), NaOH / CH$_3$OH → Polyvinylalcohol (PVA)

PS

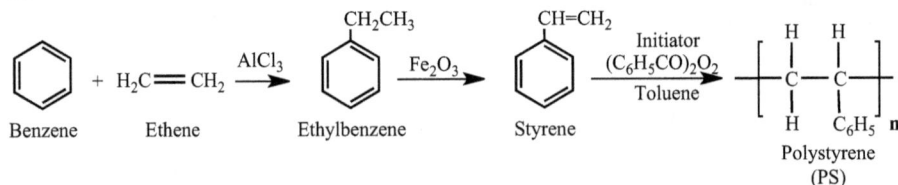

Benzene + $H_2C = CH_2$, AlCl$_3$ → Ethylbenzene, Fe$_2$O$_3$ → Styrene, Initiator $(C_6H_5CO)_2O_2$ Toluene → Polystyrene (PS)

Appendix G (Questions and Answers)

Short Questions

1. Why VC-VAc copolymer is more flexible than the PVC polymer?
   Answer: Because the presence of vinyl acetate (VAc) units in the chain decreases the intermolecular forces and thus decrease the glass transition temperature (Tg).

2. Why isotactic PP is less permeable to gas than atactic PP?
   Answer: Because isotactic PP is more crystalline and thus has less permeability.

3. Why condensation polymers are more crystalline that addition polymers?
   Answer: Due to the presence of highly polar functional group in the condensation polymers which is not the case in addition polymers.

4. Polystyrene is insoluble in methanol while its monomer is soluble. Why?
   Answer: Because of the low entropy of the polymer.

5. Why polyisobutylene does not have stereoregularity (tacticity)?
   Answer: Because it does not have asymmetric centers in its structure.

6. Polyvinyl acetate in methanol when treated with KOH precipitates out. Why?
   Answer: Because the polymer get hydrolyzed to polyvinyl alcohol which insoluble in methanol.

7. Why plasticized PVC is toxic?
   Answer: Due to the presence of plasticizers used in PVC such as phthalates, which can leach out of the material over time and cause a variety of health issues.

8. Why Teflon (polytetrafluoroethylene, PTFE) is an inert polymer?
   Answer: Due to the strong carbon-fluorine bonds, non-reactive surface, and high thermal stability.

9. Why polyurethanes have good adhesives?
   Answer: Because polyurethanes are able to bond to difficult-to-adhere-to materials, they can form strong chemical bonds with a wide range of materials, they can withstand a broad range of temperatures, and they have ability to bond to difficult-to-adhere-to materials.

10. Halogen-containing polymers should not be incinerated. Why?
    Answer: Halogen-containing polymers should not be incinerated because burning them can release toxic and harmful gases, such as hydrogen chloride (HCl). This gas is corrosive and can be hazardous to human health and the environment.

11. Why head-to-tail configuration is generally considered better than head-to-head configuration?
    Answer: Head-to-tail configuration results in a more regular and orderly structure, which leads to increased polymer stability, but Head-to-head configurations create more irregular and less organized chains. So the head-to-head because it results in more stable, stronger, and more efficient polymers with superior physical properties.

12. Why is DVB added to the polymerization of styrene?
    Answer: DVB (divinylbenzene) is added to the polymerization of styrene to serve as a crosslinking agent. DVB helps form a network structure by linking the individual polymer chains together.

13. Why polyethylene terephthalate is stiffer than polybutylene terephthalate?
    Answer: Polyethylene terephthalate (PET) is generally stiffer than polybutylene terephthalate (PBT) due to the differences in their molecular structures. PET shorter molecular structure leads to a more crystalline and tightly packed polymer, which increases stiffness, while the longer butylene chain in PBT allows for more flexibility and reduced stiffness.

14. Why LDPE polymer is a branched?
    Answer: Low-density polyethylene (LDPE) is a branched polymer because it is produced through free-radical polymerization, which involves the random initiation of polymer chains at various points along the monomer, that leads to the formation of polymer chains with various side branches, rather than a highly linear structure.

15. Why is polyethylene terephthalate (PET) less flexible than polybutylene terephthalate (PBT)?
    Answer: PET is less flexible than PBT because the shorter ethylene glycol segments in PET lead to more rigid and tightly packed polymer chains, while the longer butylene glycol units in PBT provide more flexibility due to increased chain mobility and reduced crystallinity.

16. Why the glass transition temperature (*Tg*) and melting temperature (*Tm*) of poly-vinyl chloride (PVC) and polystyrene (PS) are high?
    Answer: The high *Tg* and *Tm* of PVC and PS are due to the rigidity of their polymer backbones. The presence of bulky groups (phenyl group in PS), and the intermolecular forces (dipole-dipole in PVC and π-π stacking in PS) prevent the chains from moving freely and make it more difficult to achieve the thermal transitions associated with *Tg* and *Tm*, thereby raising their values.

17. What role do catalysts have in polymerization?
    Answer: Catalysts are chemicals that accelerate the polymerization process without being consumed in the reaction.

18. What are the differences between addition and condensation polymerization?
    Answer: In addition polymerization, monomers have unsaturated bonds that react to form a polymer chain without the loss of any atoms. In condensation polymerization, monomers have two or more reactive groups that react to form polymers, with the elimination of small molecules like water.

19. Why is the glass transition temperature (*Tg*) in polymers important?
    Answer: *Tg* is an important factor in determining a polymer's application, as it influences its performance at various temperatures.

20. What is the difference between thermoplastic and thermosetting polymers?
    Answer: Thermoplastic polymers are polymers that soften when heated and can be molded or shaped. Thermosetting polymers are polymers that, once cured or hardened, cannot be re-melted or reshaped.

MCQ Type Questions

1. Which one of the following molecules belong to polymers:
    a. Polyethylene   b. Glucose   c. Acetone
    Answer: (a)

2. Mixing of two polymer gives:
    a. Polymer Blend   b. Polymer Alloy   c. Polymerization
    Answer: (a)

3. Which one of the following is polar polymer:
    a. Polyethylene   b. Nylon 6,6   c. Polystyrene
    Answer: (b)

4. Which one of the following is nonpolar polymer:
   a. Polyethylene   b. Nylon 6,6   c. Polyvinyl Alcohol (PVA)
   Answer: (a)

5. Plasticizers are:
   a. Substances that make polymers harder
   b. Substances that prevent polymer degradation
   c. Substances that make polymers more flexible
   Answer: (c)

6. Solution polymerization is carried out in the following solvents:
   a. Water   b. Toluene   c. Acetone
   Answer: (b)

7. Solution polymerization carried out in carbon tetrachloride(CCl_4) is known as:
   a. Radical polymerization
   b. Condensation polymerization
   c. Chain-growth polymerization
   Answer: (a)

8. Which one of the following is initiator for suspension polymerization:
   a. Benzoyl peroxide
   b. Sodium hydroxide
   c. Ammonium persulfate
   Answer: (a)

9. Which one of the following polymers is a branched chain polymer:
   a. Polyethylene
   b. Polypropylene
   c. Low-density polyethylene
   Answer: (c)

10. Nylon 6 is made from:
    a. Caprolactam
    b. Hexamethylenediamine and adipic acid
    c. Terephthalic acid and ethylene glycol
    Answer: (a)

11. Nylon 6,6 is made from:
    a. Caprolactam and Hexamethylenediamine
    b. Hexamethylenediamine and Adipic acid
    c. Terephthalic acid and Ethylene glycol
    Answer: (b)

12. Nylon 6,12 is made from:
    a. Hexamethylenediamine and Dodecanedioic acid
    b. Caprolactam and Sebacic acid
    c. Hexamethylenediamine and Adipic acid
    Answer: (a)

13. Which of the following polymers will not dissolve in a solvent or melt on heating:
    a. Polyethylene   b. Phenolic resin   c. Polyvinyl chloride
    Answer: (b)

14. Tetrafluoroethylene is the monomer of:
    a. Polytetrafluoroethylene
    b. Polyvinyl chloride
    c. Polystyrene
    Answer: (a)

15. Which of the following method is not used to determine Tg:
    a. Differential Scanning Calorimetry
    b. Dynamic Mechanical Analysis
    c. X-ray diffraction
    Answer: (c)

16. Which of the following polymers are highly crystalline:
    a. High-density polyethylene
    b. Polystyrene
    c. Polyvinyl chloride
    Answer: (a)

17. The property of a homopolymer can be improved by:
    a. Blending with another polymer
    b. Increasing its molecular weight through crosslinking
    c. Adding plasticizers
    Answer: (all)

18. Which one of the following polymers is a thermoset:
    a. Polyethylene   b. Epoxy resin   c. Polystyrene
    Answer: (b)

19. Which of the following polymers is commonly used in the production of bullet-proof vests?
    a. Nylon 6,6   b. Kevlar   c. HDPE
    Answer: (b)

20. Which of the following is a characteristic property of elastomers?
    a. High rigidity   b. Elasticity and flexibility   c. Brittleness
    Answer: (b)

Fill in the Blanks Type Questions

1. The process of _____ involves the joining of monomers through covalent bonds to form a long chain polymer.
   Answer: Polymerization

2. _____ consist of a single type of monomer unit repeated throughout the polymer chain.
   Answer: Homopolymers

3. _____ polymerization is a type of polymerization in which the monomer units link together without the loss of any small molecules, typically resulting in a chain growth process.
   Answer: Addition

4. _____ polymers have a highly branched or cross-linked structure, giving them strength and rigidity, such as in materials like epoxy or Bakelite.
   Answer: Thermosetting

5. Liquid crystals show_____behavior.
   Answer: Mesomorphic

6. Most polymers can be _____and _____ identified by IR spectroscopy.
   Answer: Quantitatively/qualitatively

7. When two different monomer are polymerized together the process is known as_____.
   Answer: Copolymerization

8. Talc, fiberglass, and carbon black products are polymer additives known as _____.
   Answer: Fillers

9. PAN and PVC have strong intermolecular forces known as_____.
   Answer: Dipole-dipole forces

10. Filler material containing-polymers with improved strength is called _____.
    Answer: Reinforced plastics

11. _____polystyrene has all phenyl groups on the same side of the chain backbone.
    Answer: Isotactic

12. Polymer spherulites can be examined by polarized light microscopy and _____.
    Answer: Scanning Electron Microscopy

13. The breakdown of polymers by means of microorganisms is known as _____.
    Answer: Biodegradation

14. Thermally stable polyamide polymers are known as _____.
    Answer: Kevlar

15. The polysulfide elastomers trade name is _____.
    Answer: Thiokol

16. The preparation of tailor-made copolymers are made by _____.
    Answer: Living polymerization

17. The reinforced plastic is often called _____.
    Answer: Composite

18. The polyurethane synthetic fibers are known as _____.
    Answer: Spandex or Lycra

19. Pyrolyzed polyacrylonitrile (PAN) polymer is an example of _____polymer.
    Answer: Ladder

20. The trans 1,4-polyisoprene is known as _____.
    Answer: Gutta percha (natural rubber)

Index

https://doi.org/10.1515/9783111585734-013

www.ingramcontent.com/pod-product-compliance
Lightning Source LLC
Chambersburg PA
CBHW061416210326
41598CB00035B/6232